Biomarkers and Mental Illness

Paul C. Guest

Biomarkers and Mental Illness

It's Not All in the Mind

Copernicus Books is a brand of Springer

Paul C. Guest
Charlesworth House
Debden, Essex, UK

ISBN 978-3-319-83437-5   ISBN 978-3-319-46088-8   (eBook)
DOI 10.1007/978-3-319-46088-8

Printed on acid-free paper

This Copernicus imprint is published by Springer Nature
The registered company is Springer International Publishing AG
The registered company address is: Gewerbestrasse 11, 6330 Cham, Switzerland

Preface

A 26-year-old female sees a psychiatrist because she has developed strange ideas and convictions. She believes that the world will end soon and she is the female embodiment of the messiah who can save everyone. She also believes that she receives advice from the president of the USA through the television on how to go about this task. The psychiatrist diagnoses her with schizophrenia and, as appropriate, treats her with an antipsychotic drug called quetiapine. After several weeks, the delusions and hallucinations disappear. However, there is now another problem. She slows down, talks less and believes that there is no hope and no future. She sees the psychiatrist again and he diagnoses her with negative symptoms of schizophrenia and possible depression. As a result, he treats her with an antidepressant called fluoxetine on top of the antipsychotic medication. After 1 week, the patient becomes agitated and her messianic beliefs begin to re-emerge. Also, there is now something new—she becomes sexually uninhibited and begins picking up men in bars. The psychiatrist changes the diagnosis to mania, which is basically the upwards part of the cycle of manic depression disorder. Consequently, he stops the antidepressant treatment and gives her a mood stabilizer known as valproate. After 3 weeks there is an almost complete recovery.

There are millions of cases around the world each year such as this one. The difficulties associated with psychiatric diagnosis stem from the fact that it is still based on fuzzy concepts. This fuzziness is not only due to a broad overlap of symptoms across the various psychiatric disorders, it is likely that the underlying biological causes of these conditions overlap as well. For example, the disorder that we know as schizophrenia is likely to consist of at least five separate diseases, each of which may require a different treatment. Given the complexities and difficulties surrounding current diagnoses, the question arises—how can the process be improved?

The answer: biomarkers.

I have written this book on the emerging use of biomarkers in the study of psychiatric diseases for a broad range of people including researchers, clinicians, psychiatrists, university students and even those whose lives are affected in some way by a psychiatric illness. The latter category is not trivial since a staggering one in three people worldwide show the criteria for at least one psychiatric disorder at some point in their lifetime. The book lays out, in accessible language, the history of psychiatric research, the current state of the art in psychiatric practice, the systems affected in psychiatric illnesses, the whole body nature of psychiatric illnesses and the impact that this is having on emerging biomarker discoveries. It also gives descriptions of the major specific psychiatric disorders and the special challenges that surround the diagnosis and treatment of each one of these. The main concept behind this book that the reader should look for is that the brain does not work alone. Mood and behaviour actually result from integration of signals between the mind and the body, and many of these signals are borne by the bloodstream. This is important as this factor makes it possible to develop simple blood tests for diagnoses of these complex disorders. The final chapter drives home the way in which we can change the paradigm of how we treat patients with psychiatric disorders by incorporating biomarkers into clinical practice and even into the drug development pipeline. The ultimate goal is to incorporate personalized medicine approaches into these processes to help move psychiatric medicine into the twenty-first century.

I sincerely hope that the reader will find the contents of this book interesting on a subject that really matters and affects us all in some way.

Debden, Essex, UK   Paul C. Guest

Contents

Part I
A Brief History of Biomarkers and Mental Illness

Chapter 1
Psychiatric Disorders as "Whole Body" Diseases

In the last decades of the nineteenth century "insanity" was postulated to be a set of discrete mental conditions, which were thought to exist alongside other diseases that are widely recognized by general medicine. However, anatomical studies which focused on the search for pathological processes in the brains of psychiatric patients found little or no evidence for a link with the common clinical signs or symptoms. Nevertheless, the new disciplines of endocrinology, immunology and biochemistry arrived on the scene, and began to offer new insights that pointed towards involvement of the whole body in the precipitation and course of psychiatric disorders. The assumption behind this research was that the brain was not separate from most of the physiological currents that surged throughout other organs and peripheral systems of the body. Thus, the source of insanity might reside not just in the brain, but elsewhere in the body. This chapter will briefly cover the history of physiological models of psychopathology with respect to what we now call schizophrenia, the major mood disorders and other psychiatric Illnesses. Over the past 60 years, most historical narratives have focussed on neuroanatomy, neurotransmitter systems and the genome. There is now an emerging concept that psychiatric disorders can be precipitated by various factors external to the brain, including exposure to certain drugs, infections, nutrient deficits, physical and emotional stress, and by other diseases such as autoimmune conditions and metabolic disorders. With the emergence of proteomic biomarker research at the start of the twenty-first century, whole body causes of psychiatric diseases are once again the prime suspect.

A Brief History of Psychiatric Disease: The Curse of the Gods?

In prehistoric times, many individuals believed that mental diseases came from magical beings that disrupted the minds of unsuspecting and innocent victims (Fig. 1.1). Ancient shamans used spells and rituals in an attempt to cure the afflicted

© Springer International Publishing AG 2017
P.C. Guest, *Biomarkers and Mental Illness*,
DOI 10.1007/978-3-319-46088-8_1

Fig. 1.1 Many people in
ancient times believed that
mental illnesses were
caused by the gods

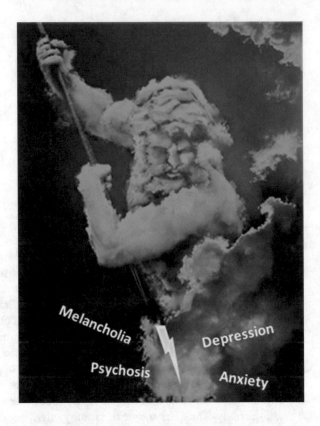

of their mental illnesses This often took the form of an exorcism, which involved the
shaman attempting to drive the invading spirit away from the body. In other cases, a
crude form of brain surgery was used which was called trepanation. This involved
drilling a hole through the skull to allow the spirits trapped inside the skull an easy
escape route to the outside world. Skulls with these holes have been found in Europe
and South America, dating back some 10,000 years (Fig. 1.2). Amazingly, the technique is still in use today for conditions such as epidural and subdural hematomas,
and for allowing surgical access for particular procedures such as intracranial pressure monitoring.

The ancient Egyptians brought about many changes in the treatment of the mentally ill. However, they still regarded these conditions as resulting from magic or
that they were brought about by the gods. For example, they believed that normal
mental health required a strong interaction with the *khat* (the body), the *ka* (the
guardian spirit) and the *ba* (a bird symbolising the connection with heaven).
Concepts similar to these forces can be found in many cultures, such as chi or qi in
china, *mana* in Hawaiian culture, *lüng* in Tibetan Buddhism, as well as ideas in
popular culture (e.g. "The Force" in the *Star Wars* films and books). The obsession
of the Egyptians with life after death revolved around the belief that the well-being
of the mind and soul was critical for overall health of each person. For this reason,

Fig. 1.2 Ancient skull showing a skull subjected to trepanation—a drilling process used to let the bad spirits out

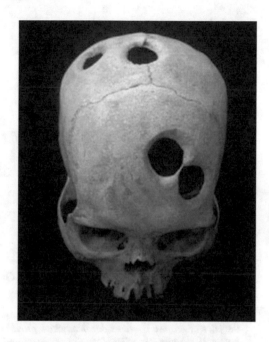

they focussed on the health of the soul and they carried out specific purposeful treatments for erratic behaviour as early as 5000 years ago. Typical cures for those suffering from any form of insanity included treatment of the afflicted with opium, performance of ritual prayers to the gods, along with dream interpretation to discover the source of the illness.

Is It All in the Blood or Other Body Fluids?

The ancient Greeks were probably the first culture in known history to overcome the belief systems that diseases were caused by supernatural events. In fact, they came to realize that afflictions of the mind did not differ from those of the body. Following on from this, they developed a system which saw all sicknesses as being due to natural phenomena and the cause may reside in the blood and other fluids (humours).

For thousands of years, the blood has been used as a source of information on illness and health in human beings. Ancient Greek doctors believed that most illness was caused by an imbalance in the four body humours known as black bile (earth), yellow bile (fire), phlegm (water) and blood (Air) (Fig. 1.3). More specifically, they believed that fever was caused by an imbalance of the yellow bile and they would therefore try to increase the opposite humour phlegm (water), for example by immersing the afflicted in a cold bath. In the case of mental illnesses, an excess of phlegm (water) was thought to render a person emotionally unresponsive. The father of medicine, Hippocrates, believed that an excess of black bile led to irrational

Fig. 1.3 Ancient Greek doctors believed that most illnesses are caused by an imbalance in the four body humours known as black bile (earth), yellow bile (fire), phlegm (water) and blood (air)

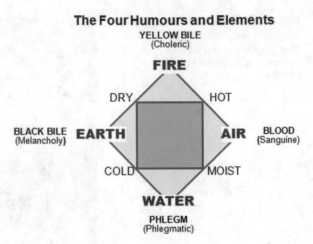

thinking or depression. So as a response, the Greeks would use bleeding and purging to rebalance the blood and black bile.

> *Aristotle, who used the image of wine to expose the nature of black bile. Black bile, just like the juice of grapes, contains pneuma, which provokes hypochondriac diseases like melancholia. Black bile like wine is prone to ferment and produce an alternation of depression and anger....* (From Linet's, *The History of Melancholy*)

Again, these four humours have analogues in other cultures. For example, the Chinese concept of qi revolves around five elements: wood, fire, earth, metal and air. These are thought to function in a cyclical sequence that represents the natural flow of qi. Furthermore, these elements are thought to represent many qualities. In terms of emotion, wood represents anger versus patience, fire is hate versus joy, earth corresponds to anxiety vs empathy, metal equals grief versus courage and water represents fear versus calmness. The Chinese developed a number of exercises, such as qi gong and tai chi, to help the qi keep flowing to maintain physical and mental health and to rebalance these elements during sickness.

> *Author's note: I have tried both of these at various times and found them to be both relaxing and stimulating. In the case of qigong, I managed to improve my ability to stand still like a tree (with arms outstretched at a 90 degree angle from the body) from 5 minutes to 20 minutes.*

What Went Wrong in the Middle Ages?

By the time the Middle Ages came around, such progressive ideas about the cause of psychiatric disorders were discarded. People who were mentally ill were now treated with fear, disgust and shame. Also, hospitals known as asylums began to emerge in Egypt and Iraq around the thirteenth century and these were not very hospitable places. These were used to isolate the mentally ill or the socially ostracized from society, with purpose of removing them as a problem. In fact, asylums did not offer any form of treatment to help the afflicted to reintegrate into society. Europe's oldest asylum was the Bethlem Royal Hospital of London, which became known as the infamous Bedlam. Bedlam began housing the mentally ill in 1403. The first asylum in the Colonial USA was built in Williamsburg Virginia around the year 1773. Pictures from some of these institutes revealed that the "patients" were often bound with rope or chains to beds or walls, or restrained in straitjackets.

With the emergence of techniques in experimental physiology and medicine in the middle of the nineteenth century, investigations into the biological nature of mental illness began to re-emerge. In an amazing venture well before its time, many studies were conducted which attempted to determine whether or not there were physical characteristics in blood that could be used to distinguish the diseased state from the normal one. In addition, questions arose as to whether blood characteristics could be used to determine whether insanity was a continuum of one disease or a variety of discrete disease entities that could be identified and sorted into different clinical categories. According to Richard Noll in his paper "The blood of the Insane" in the *History of Psychiatry*, these investigations comprised four phases based on different concepts (summarized below).

Phase 1

In 1854, the first microscopic investigations of blood cells from psychiatric patients were carried out in a Scottish asylum by W. Lauder Lindsey. He basically used a low power microscope to count the numbers of different blood cells in samples from the patients in comparison to those of his staff members. However, this investigation did not yield any earth shattering results as Lindsey found no differences associated with different types of mental illness or between the patients and the staff members. However, approximately 30 years later, S. Rutherford Macphail reviewed a number of subsequent related studies and came to the conclusion, albeit a tentative one, that there was an overall "deficiency of corpuscular richness of the blood" in the early stages of mental illness. This meant that he thought there may be a lower number of red blood cells in psychiatric patients. Similar blood studies of patients at various stages of insanity continued up to the 1920s with mixed results and no firm conclusions.

Phase 2

As the new field of endocrinology began to emerge in the 1890s, the scientists of that time jumped on the band wagon and began to study blood to detect and measure "inner secretions" in an attempt to increase our understanding of many medical conditions, including mental illnesses. Thus, this endocrinological approach was adopted by the first modern biological psychiatrists who were looking for a new approach in understanding the physical causes of psychiatric disorders. This was driven by the assumption that if an over or under production of "inner secretions" could lead to diseases such as diabetes (this secretion was identified as insulin in 1921 by Frederick Banting and Charles Best), which were accepted as diseases of a physical nature, then the same might be true of one or more of the psychiatric disorders. One German psychiatrist in particular, Emil Kraepelin (1856–1926), followed this approach doggedly and eventually developed the concept that severe psychotic disorders (he first called these "dementia praecox") were the result of a persistent whole body metabolic disease process which led to effects on the brain in the later stages, resulting in chronic mental "deterioration". It was from these ideas that Kraepelin became regarded by historians as a central figure in the history of modern psychiatry. Then, in 1908, a Swiss psychiatrist named Eugen Bleuler (1857–1939) proposed the new term "schizophrenia" as an expanded version of Kraepelin's dementia praecox. However, schizophrenia was meant to have a more favourable disease outcome. Nevertheless, only the term schizophrenia and not the original disease concept of Bleuler are accepted in the present day. For more than 100 years changes in the function of hormones and hormonal abnormalities in various psychiatric disorders have been detected and this remains a frequent finding today. The endocrinological paradigm also provided a direct link to the discovery of acetylcholine as the first neurotransmitter, by German biochemist Otto Loewi in 1921. We will see later how an upset in the balance of neurotransmitters plays a key role in the development of various psychiatric conditions.

Phase 3

In 1906, an immunoserodiagnostic paradigm emerged in psychiatric research. This was basically looking at immune factors in blood to help distinguish psychiatric patients from "normal" healthy subjects. This followed the development of the Wasserman reaction test for neurosyphilis, which was considered a breakthrough in biological psychiatry. This was important as it actually marked the first occasion in which a diagnostic blood test was used to identify a specific mental illness, commonly observed in asylums and known as "general paralysis of the insane". Following this, two German psychiatrists injected cobra venom into dementia praecox and manic-depression (bipolar disorder) patients. This showed that some patients had reactions to the toxin compared to healthy control subjects. However, these results were never replicated and their findings were later refuted.

> Author's note: replication of results is paramount in all scientific studies. If they cannot be replicated, they never really happened.

This work was followed by a more influential immunoserodiagnostic test produced by a Swiss biochemist named Emil Abderhalden (1877–1950) and one developed by the German psychiatrist August Fauser (1856–1938). Both of these tests were aimed at diagnosis of dementia praecox and manic-depressive conditions in comparison to healthy subjects. Again, none of these findings could be replicated and these tests were also cast into doubt. Nevertheless, over the last 100 years or so, changes in immune function and inflammation which can be detected in the blood have been linked to various psychiatric disorders through many lines of evidence and there remains no doubt about their involvement (at least at some level) in psychiatric disorders.

Phase 4

The development of blood tests for psychiatric illnesses in the twenty-first century has been attempted based on breakthroughs in the fields of genomics (the study of genes) and proteomics (the study of proteins). The way it works is that all cells of a particular organism contain the same genes (the genome). However, only specific sets of genes are expressed as proteins (the proteome) and this depends on specific factors such as the type of cell, the developmental stage of the organism or the presence of external factors like disease. As far as the genes are concerned, heredity has been shown to play in role in the development of different psychiatric disorders to varying extents. However, efforts at developing genetic-based diagnostic tests for most medical diseases including psychiatric illnesses have met with disappointment. Despite 20 years of extensive efforts, no single gene, or combination of genes, have been identified that could be linked to an increased probability of an individual developing any psychiatric disorder. Likewise, no studies carried out thus far have implicated a single protein as being causative in any of the major mental illnesses. However, a number of recent studies have suggested that such an effect may occur through effects on protein networks. This means that not one but multiple proteins may be affected in specific disorders. Therefore, there is considerable hope that specific patterns of altered proteins (proteomic fingerprints) can be found and linked to different psychiatric disorders (see later chapters).

Famous People with Psychiatric Disorders

The commonality of mental disorders is perhaps best illustrated by the number of famous individuals, past and present, who have been afflicted. Throughout history and contemporary times, many people with so called mental illnesses have contributed immensely to society and human culture. Despite their accomplishments, many faced stigma within their lives. One reason why many of these individuals may not be known to the reader is due to the fact that many of them have learned to cope and even conquer their demons. On the list below are many individuals who have made amazing contributions to the arts, sciences or the world of politics. There are of course many others but space limitations prohibit a more comprehensive list.

SCIENTISTS
Isaac Newton (25 December 1642–20 March 1727)
Field: Mathematician/scientist of the seventeenth Century, responsible for many
    scientific discoveries
Condition: suspected bipolar disorder, schizophrenia.
Nikola Tesla (10 July 1856–7 January 1943)
Field: physicist/inventor
Condition: suspected obsessive compulsive disorder
Albert Einstein (14 March 1879–18 April 1955)
Field: Physicist (Nobel Prize winner)
Condition: suspected Asperger syndrome
John Nash (June 13, 1928–May 23, 2015)
Field: Mathematician (Nobel Prize winner)
Condition: schizophrenia

ARTS/LITERATURE
Leo Tolstoy (9 September 1828–20 November 1910)
Field: Writer
Condition: depression, hypochondriasis, alcoholism and substance abuse
Vincent Van Gogh (30 March 1853–29 July 1890)
Field: Painter/artist
Condition: suspected major depressive disorder or bipolar disorder (death by
    suicide)
Louis Wain (5 August 1860–4 July 1939)
Field: Painter/artist
Condition: schizophrenia
Virginia Woolf (25 January 1882–28 March 1941)
Field: Novelist
Condition: bipolar disorder (death by suicide)
Ernest Hemingway (July 21, 1899–July 2, 1961)
Field: Novelist/Short-story writer/journalist (Pulitzer Prize and Nobel Prize winner)
Condition: depression (death by suicide)
Jack Kerouac (March 12, 1922–October 21, 1969)
Field: Novelist
Condition: dementia praecox (older term for schizophrenia)

MUSICIANS/COMPOSERS
Wolfgang Amadeus Mozart (27 January 1756–5 December 1791)
Field: Composer
Condition: suspected autism spectrum disorder (Asperger syndrome)
Ludwig van Beethoven (baptized 17 December 1770–26 March 1827)
Field: Composer
Condition: bipolar disorder
Brian Wilson (born June 20, 1942)
Field: Songwriter/musician
Condition: schizophrenia (although diagnosis later retracted)

Janis Joplin (January 19, 1943–October 4, 1970)
Field: Singer/songwriter was a blues-rock singer-songwriter who rose to promi-
   nence during the Condition: addictions (died from drug overdose)
Syd Barrett (6 January 1946–7 July 2006)
Field: Songwriter/musician
Condition: suspected schizophrenia
Billy Joel (born May 9, 1949)
Field: Songwriter/Musician
Condition: depression
Karen Carpenter (March 2, 1950–February 4, 1983)
Field: Singer/musician
Condition: anorexia nervosa (died from cardiac arrest)

POLITICIANS/ROYALS
George III (4 June 1738–29 January 1820)
Post: King of England
Condition: porphyria (an enzymatic disorder that affects the nervous system, result-
   ing in mental disturbances, including hallucinations, depression, anxiety and
   paranoia)
Abraham Lincoln (12 February 1809–15 April 1865)
Post: President of the United States
Condition: Suspected depression
Winston Churchill (30 November 1874–24 January 1965)
Post: Prime Minister of Great Britain (also officer in the British army, a historian,
   impressionist painter and Nobel Prize-winning writer)
Condition: suspected bipolar disorder
Diana, Princess of Wales (1 July 1961–31 August 1997)
Post: Princess of Wales
Condition: suspected depression

PERFORMING ARTS
Harrison Ford (born July 13, 1942)
Field: Actor
Condition: depression
Robin Williams (July 21, 1951–August 11, 2014)
Field: Actor
Condition: depression
Dan Akroyd (born July 1, 1952)
Field: Actor
Condition: Asperger syndrome and Tourette syndrome
Carrie Fisher (born October 21, 1956)
Field: Actress
Condition: bipolar disorder
Daryl Hannah (born December 3, 1960)
Field: Actress
Condition: autism

Owen Wilson (born November 18, 1968)
Field: Actor
Condition: depression and drug addiction
Heath Ledger (4 April 1979–22 January 2008)
Field: Actor
Condition: suspected depression, anxiety and drug addition

Is It All in the Mind?

In order to understand the aetiology of a disease which has traditionally been regarded as a disorder of the mind, it is logical to begin with a study of the brain. The brain is basically an integrated circuit board with different regions responsible for carrying out different functions. Like the four humours and the five elements of qi, illness results when the function of these different regions become imbalanced due to over- or under stimulation. In order to visualize how this happens, it is necessary to know that these different regions are connected by neuronal fibres as if by electrical wiring, with both positive and negative inputs. Thus, over stimulation of a positive input or under stimulation of a negative input could lead to excessive excitation, and *vice versa*. The following sections will describe the basic brain regions and the roles that these play in the regulation of cognition, mood and behaviour. These sections will also cover the connective wiring which is regulated by the activity of small molecules known as neurotransmitters.

It is not likely that individual brain regions perform specific tasks. It is more likely that two or more brain regions work together in connected pathways in order to achieve this. The following sections will describe the process from the ground up. Thus, we will begin with the neurons themselves and the neurotransmitters that they use for transmission of the signals and then move to the different brain regions. Finally we will put this all together and describe how all of this can lead to either a balanced or an imbalanced brain when it comes to mood, behaviour, cognition and other mental functions.

The Basics: Neurotransmission

First we will begin with the basics of how neurotransmission occurs in the brain. Neurons are unusual cells in the body as they are shaped like trees with a bump in the middle, where the cell body and nucleus reside. From the cell body, the dendrites emerge like the branches of a tree in one direction and a single axon emerges like the thicker trunk from the other (Fig. 1.4). Dendrites receive signals from other neurons and pass these down to the cell body and through the axon. The axon is enclosed in a fatty tissue called myelin, which acts as sort of an insulator to ensure efficient signal connection (a lot like the case of electrical wiring which is often

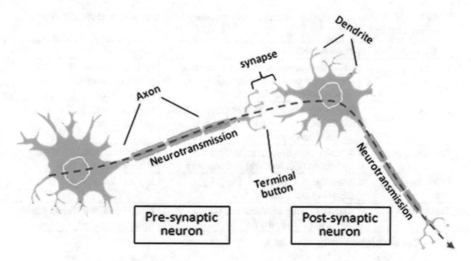

Fig. 1.4 Neurotransmission from a presynaptic neuron to a postsynaptic neuron

Fig. 1.5 *A synapse.* The *arrow* shows the direction of neurotransmission. *NT* neurotransmitter

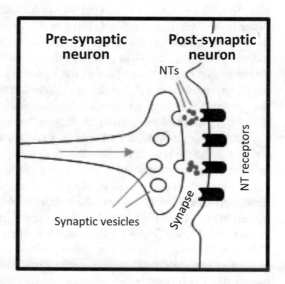

encased in plastic or rubber insulation). The end of the axon is comprised of a bulbous projection called a terminal button which contains a lot of little spheres called synaptic vesicles. Each vesicle contains a neurotransmitter such as dopamine, serotonin, glutamate, GABA, acetylcholine or neuroepinephrine, which are essentially chemical messengers. These vesicles can move and then fuse with the terminal button in the process of releasing the neurotransmitters into a space called the synapse (Fig. 1.5). The neurotransmitters then cross this gap and bind to specific receptors embedded in the membranes of the dendrites of the next neuron to continue the

Table 1.1 Neurotransmitters, their functions and associated diseases which occur due to over- or under-production

Neurotransmitter	Normal function	Disease
Dopamine	Controls movement, posture and mood, and plays a central role in positive reinforcement and dependency	Parkinson's disease, ADHD[a], pain, drug addiction, schizophrenia
GABA	Inhibitory neurotransmitter that regulates motor control, vision and many other cortical functions	Epilepsy, anxiety disorders, schizophrenia
Glutamate	A major excitatory neurotransmitter associated with learning and memory	Alzheimer's disease, stroke, schizophrenia, autism
Norepinephrine	Attentiveness, emotions, sleeping, dreaming and learning	Bipolar disorder, depression
Serotonin	Regulates body temperature, sleep, mood, appetite and pain	Depression, suicide, impulsive behaviour and aggression

[a]Attention deficit hyperactivity disorder

transmission of the signal. This sounds like a lot of steps but the whole thing actually takes around 5 ms. In fact, the speed of a neurotransmitter crossing the synaptic cleft has been estimated at around 440 kph (270 mph). This is almost ten times faster than Usain Bolt running at top speed. The neuron sending the message from its terminal buttons is called the presynaptic neuron and the receiving neuron is called—you guessed it—the postsynaptic neuron. What happens to neurotransmitters after they have finished passing the signal to receptors on the postsynaptic neuron? The answer is one of two things: (1) the neurotransmitter is taken back up into the presynaptic neuron for repackaging into synaptic vesicles for release at a later time; or (2) it is broken down by an enzyme arising from the postsynaptic neuron.

When a neurotransmitter stimulates a postsynaptic neuron, changes occur that are specific to the type of neurotransmitter. Some neurotransmitters cause channels to open up in the postsynaptic axon and this allows ions that are floating around outside the cell to enter it. Depending on which ions and how many are involved, these can cause an electrical impulse to travel down the axon. The basic neurotransmitters which have been implicated in mental illnesses are dopamine, gamma-aminobutyric acid (GABA), glutamate, norepinephrine and serotonin (Table 1.1). This is mainly caused by over or under production of these molecules.

One of the main theories surrounding psychiatric disorders is that the disease occurs due to an impairment or loss of connectivity. That is, brain regions which normally communicate in the processing of such complex functions as perception, cognition, mood and emotions are no longer connected in optimal manner.

The amygdala is the fear centre of the brain which activates our instinctive fight-or-flight response to dangerous situations. It may also be involved in learning to fear and not to fear certain events. The prefrontal cortex is where control of executive functions such as judgment, decision making and problem solving take place. Different parts of this brain region are also involved in the use of short term memories and the retrieval of long term ones. Interestingly, the prefrontal cortex can temper the amygdala response during stressful events. The anterior cingulate cortex regulates attentiveness, motivation and emotional reactions. The hippocampus is involved in creation and storage of memories. The cerebellum has been mainly known for its role in regulated gross and fine motor movement. However, recent studies indicate that it also plays a role in regulation of higher functions such as mood and emotions. These and other brain regions will be discussed in more detail as their dysfunction relates to various psychiatric disorders.

The Disconnectivity Theory of Mental Disorders

After approximately 30 years of intensive studies of psychiatric disorders, no single site of pathology has been identified as a causal factor. Instead, the findings have suggested that subtle changes occur in multiple brain areas, which could mean that it is all about disordered interactions between local and distal neural centres rather than damage to any individual brain region (Fig. 1.6). In fact, communication and integration of information between brain regions is known to be required for healthy brain function. As mentioned above, it is known that the prefrontal cortex region of the brain tempers the activity of the amygdala. This is basically achieved by a connection from the prefrontal cortex which tells the amygdala not to overreact. Excitatory outputs project to the basal nucleus of the amygdala, which then excites the central nucleus to elicit a fear response. In contrast, another set of excitatory outputs from the prefrontal cortex leads to stimulation of GABA inhibitory neurons in the intercalated cell masses. Since these signals are inhibitory, this leads to inhibition of the central nucleus, dampening down the conditioned fear.

In addition to its well known role in regulation of motor functions, it is now known that the cerebellum is also involved in regulation of cognition and other higher brain functions such as mood and emotions. Structural proof for this comes from the finding that the cerebellum receives nerve inputs from several brain areas such as the prefrontal cortex, which regulates higher brain functions. Likewise,

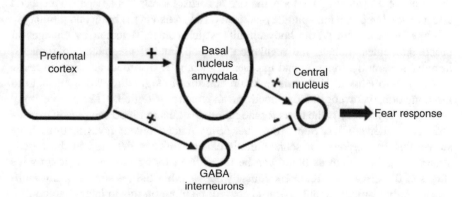

Fig. 1.6 Diagram showing prefrontal cortex–amygdala connectivity, leading to positive or negative regulation of the fear response

nerves from the cerebellum have been traced through a circuitous route to both motor and non-motor areas of the prefrontal cortex. Furthermore, functional evidence for a role of the cerebellum in higher brain function has been demonstrated by a technique called magnetic resonance imaging (MRI) (more on this later), which showed that neuronal activity in the cerebellum correlated with changes in activity in other brain regions including the prefrontal cortex.

What Can We Learn About Brain Function by Studying the Serum?

One question I always receive when giving lectures is: "how can you study the brain by looking at the blood?" The answer is simple and is always something along the lines of: "the brain does not work alone." Blood serum and plasma have been used increasingly in biomarker studies of psychiatric disorders for more than a decade. The rationale for this comes from the fact that psychiatric disorders are whole body diseases (just like the Ancient Greeks believed). The fields of endocrinology, immunology and biochemistry have shown that the brain is integrated in fundamental bodily functions, which is also reflected in changes in the composition of blood proteins and other bioactive molecules. The disciplines of endocrinology, immunology and biochemistry have revealed that the brain is holistically integrated in the most fundamental biological functions of the body, which is also reflected in changes in the composition of the blood. One of the best examples is the fight or flight reflex. The reaction begins with the often unconscious perception of danger mediated by the amygdala and additional afferent regions of the limbic system, projecting to the hypothalamus via the stria terminalis, triggering the release of corticotrophin releasing factor (CRF). This results in increased secretion of adrenocorticotrophic hormone (ACTH) from the pituitary into the bloodstream. ACTH circulates throughout the body and acts on the adrenal glands to produce and release the hormone cortisol. This causes an increase in blood pressure and blood glucose, in preparation for the muscular actions required in the response. Completing the loop, the rise in cortisol levels has a negative feedback effect on the hypothalamic–pituitary–adrenal (HPA) axis via the brain and pituitary.

Other well known whole body circuits which can be detected by changes in blood molecules include the regulation of food intake, pregnancy in females, immune system dysfunction and in cases of metabolic disorders and insulin resistance. Impairments in HPA axis function and insulin signalling can contribute to inflammation, neurological dysfunctions and memory deficits. The whole body concept in psychiatry is the basis for the rationale of using blood samples in psychiatric biomarker studies. Supporting this development, recent investigations have shown that biomarkers for schizophrenia can be reliably detected in the bloodstream. This is critical as blood can be taken from living patients at progressive stages of the disease or treatment course whereas other tissues such as brain biopsies are only available at autopsies. There will be more on this in later chapters.

Chapter 2
Treatment of Psychiatric Disorders: Time for a Paradigm Change?

This chapter outlines the clinical need for improved diagnostic and therapeutic strategies in the field of psychiatric disorders. There is currently only a medium success rate for psychiatrists in accurate diagnosis of psychiatric patients. In addition, the response rate of patients to current medications is only mediocre once a correct diagnosis has been made. These factors can combine and thus have a negative impact on disease outcome. Furthermore, virtually no novel compounds for treatment of these disorders have entered the market over the last few decades. Instead, most new drugs are merely derivatives of the existing ones. This is due to a lack of good model systems in pharmaceutical company pipelines for the testing and development of novel drugs. As a consequence, many of the major pharmaceutical companies are deserting the field of psychiatry as a potential drug market. This chapter introduces the ideas and anticipated benefits of a shift to more individualized or personalized medicine approaches in the identification and treatment of patients with psychiatric disorders (more detailed information is given in Chap. 12). This shift will involve the use of biomarkers for better classification of patients and for use of a biomarker-related readout to guide the discovery and selection of the most appropriate drugs to give to the patients. This has been exemplified by a review of the oncology field, in which such approaches are revolutionizing patient care and survival. There is no reason to assume that this cannot also work for psychiatry.

What Is the Clinical Need?

There is currently an unmet clinical need for molecular biomarkers in studies of major psychiatric disorders to improve diagnosis and treatment options. Thus far, identification of such biomarkers has not borne fruit, most likely due to the fact that the various disorders are still classified based on long-standing out-dated diagnostic concepts used in psychiatry, and because the various disorders are notoriously heterogeneous in terms of their aetiology and symptoms. Also, the identification of

P.C. Guest, *Biomarkers and Mental Illness*, DOI 10.1007/978-3-319-46088-8_2 biomarkers for a disease that has already been categorized based on clinical symptoms may not be useful in the clinic if such categorisations are indeed inaccurate. Thus, innovative approaches for identification of biomarkers for psychiatric disorders are needed which can be used to classify at-risk patients, such as teenagers with prodromal symptoms for psychosis as well as existing patients who are likely to deteriorate to even more severe states. Many researchers and psychiatrists are now in full support of the idea to deconstruct the traditional diagnoses in favour of more empirical methods, such as classification by use of serum or plasma biomarker panels. This is not intended as a replacement for the old method but as a supplement to be co-administered for increased diagnostic accuracy. This is important since the old ways are working—just not a hundred percent of the time. In other words, there is plenty of room for improvement.

What Is a Biomarker?

Biomarkers are physical characteristics that can be measured and evaluated as an indication of a normal biological process, a disease, response to a drug or toxin, or even the history of a medical condition (Fig. 2.1). To be more precise, the FDA has defined biomarkers as "measurable characteristics that reflect physiological, pharmacological, or disease processes in animals or humans". In medical practice, a valid biomarker could be used to identify a syndrome, a treatment response or a clinical course. For practical purposes a biomarker should be measurable with excellent accuracy and reproducibility, within an acceptable time-frame and at a low cost. Ideally, a biomarker would reflect in some way the underlying nature of the disorder or the biological process affected by a drug treatment.

In psychiatric research, the term biomarker has been used for various kinds of measurements ranging from brain imaging analyses, neuropsychological tests, electrophysiological responses or older methods and techniques such as the skin flush response after chemical provocation. This is based on the idea that "normal" people

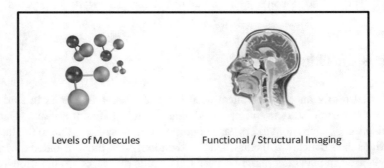

Fig. 2.1 Biomarkers are measured characteristic that can give an indication of physiological status or change

show a skin flush or reddening effect after topical application of niacin to the skin and this may be blunted in patients with schizophrenia. This chapter focuses on the use of molecules identified in peripheral blood (i.e. protein or gene expression data). Furthermore, it focuses on the major psychiatric disorders: schizophrenia, bipolar disorder and major depressive disorder.

Clinical Examples

The following section includes several real life psychiatric case studies for which the clinical outcome could have been improved by incorporation of biomarkers to help with clinical decision making. In all cases the diagnosis of the particular psychiatric condition was uncertain and often changing, and yet treatment was prescribed.

Case 1

A 17-year-old male high school student began to show increased periods of absence from his classes. In addition to this, his test scores were lower than they were previously and he seemed to have problems with attention. It was not long before he was referred to a psychologist. The psychologist diagnosed the student as having an adjustment disorder with an alteration in mood and the possible occurrence of a major depressive episode. The psychologist saw the potential causes for this as relating to the student's uncertainty about future studies as well as conflicts with his father and younger brother. Based on these observations and report histories, the psychologist prescribed treatment with an antidepressant and cognitive behavioural therapy (CBT). In CBT, the therapist tries to help the patient by helping them to choose better strategies for dealing with life's difficulties. At first, the student showed some improvement and he began to attend his classes more regularly, although his overall performance was still lower than before. Six months later, when the student began the final year of his high school, his condition began to deteriorate. During this time he stayed at home and said that he no longer wanted to go to school because he found it to be useless. He also refused to see the psychologist any more for the same reasons.

Naturally, the student's parents had no idea about what to do next. One day, the father made his son get into the car in an attempt to force him to go to his classes. However, the boy ran away and did not return home until several hours later. After this, the boy did not leave his room for several days. The family doctor urged that the student should be admitted into a psychiatric hospital, but the boy refused. A few weeks later, the boy stopped eating because he claimed that the food was poisoned with radio-activity. After this, he underwent forced admission into the psychiatric hospital (this is referred to as involuntary commitment). Over a period of several days, the patient described his beliefs that his school was a centre of evil scientists who were involved in covering up nuclear accidents and these scientists wanted to kill him because he had discovered this fact. Given this delusion and other factors mentioned above, the boy was diagnosed with schizophrenia. His doctors attempted to treat him with the antipsychotic drugs risperidone and aripiprazole but these led to no improvement. The patient finally improved after receiving treatment with olanzapine, although he still showed some problems in attention and working memory performance. Olanzapine is one of the most wildly prescribed and strongest antipsychotics for treatment of psychosis. Next, his doctors attempted a re-entry programme into the school in which the boy would resume classes the next year, although on a slightly lower attendance level. However, the patient began to show considerable weight-gain over a period of several months to the point of borderline obesity [obesity is defined as having a body mass index (BMI) over 30 kg/m^2]. This is a well known side-effect of olanzapine treatment. Nevertheless, the student did show some improvement of symptoms and the treatment with olanzapine was continued.

What could biomarkers have told us? Well, this case is a classical example on the development of schizophrenia, beginning with symptoms such as atypical mood and motivational changes, as often seen during adolescence. It is not unusual that in the prodromal phases of schizophrenia and other psychiatric illnesses that these symptoms may be attributed to the typical behaviours of adolescence. However, a number of recent scientific studies have now begn to focus on developing a means of identifying young patients who are at-risk for mental disorders such as schizophrenia and distinguishing these from milder disorders, which usually have a more limited time course. These prodromal periods are often referred to as 'at-risk mental states', which are usually identified using symptoms that have limited reliability or the presence of certain genetic risk factors. The availability of a biomarker test indicating a likely illness trajectory would have been highly beneficial here. The other factor of this case was that serious metabolic side effects occurred in the form of weight gain, due to a common side effect of antipsychotic treatment. Therefore, a biomarker test which could predict this response may be beneficial as it would allow potential add-on treatments which could potentially combat the weight gain. For example, an anti-diabetic medication could have been used in combination with the antipsychotic. Later chapters describe studies which have attempted to carry this out.

Case 2

A 26-year-old female bank manager was referred to a psychiatrist because she developed several strange ideas including that she was a reincarnation of the prophet. She also said that the future of mankind was threatened and she was insistent that everyone should listen to her warnings. She claimed that she received coded messages from the president of the United States through her television, giving her advice about the fate of the world. She was diagnosed with schizophrenia and her doctors prescribed treatment with the antipsychotic quetiapine. This had some positive effects through elimination of the delusions and hallucinations but the patient became visibly slower in movement and speech. When she did speak, the only statements she made were those referring to the absence of a future. Following this, her doctor gave a second diagnosis: 'negative symptoms of schizophrenia, possibly depression' and he prescribed an antidepressant in combination with the antipsychotic medication. After 1 week of receiving this new treatment approach the patient showed signs of renewed agitation and restarted her statements about being a prophetess, along with the statements about the end of the world. However, another behaviour began emerge. She began to show signs of sexual disinhibition and even solicited men in bars. She was then admitted to a psychiatric hospital. In the hospital, she wore excessive make-up and talked incessantly. The diagnosis was then changed to mania as part of bipolar disorder or schizoaffective disorder and she was given a mood stabilizer. This treatment had many positive effects and after 3 weeks there was an almost complete recovery.

This case is an example of a patient who has symptoms suggestive of one disorder who then appears to switch to another condition. The reasons for the misdiagnosis are most likely due to the fact that current psychiatric diagnoses are subjective in that they are based on concepts that have both clinical and biological overlaps. However, the availability of a biomarker panel that afforded more accurate diagnosis might have been used to identify the disease correctly. Theoretically, this would allow the patient to be treated with the right drug targeting the right disease for the best possible clinical outcome. This is also in line with the "personalized medicine" concept which is also discussed further in later chapters.

Case 3

A 19-year-old female college student with general good health and no psychiatric history developed auditory and visual hallucinations over a 2-week period. Her general practitioner referred her to a general mental health care facility, where she received an initial diagnosis of psychotic disorder not-otherwise-specified with a possible personality disorder. Based on this, she was prescribed treatment with olanzapine. However, the hallucinations were still present after 1 week and delusions also began to emerge. One of these was the belief that her parents were poisoning her food. She was therefore admitted to a psychiatric hospital, where several clinical diagnoses were considered. These included psychotic disorder not-otherwise-specified, schizophreniform disorder, major depressive disorder and psychotic/mood phenomena with a personality disorder. Routine neurological examinations and laboratory tests were performed but these showed no abnormalities. A gradual amelioration occurred over the next few weeks but the patient showed an inability to concentrate, an altered personality (based on reports of her parents) and mood swings. She was given a psychological test while she was in a state of partial remission which revealed that she had deficits in working memory and sustained attention, and this led her doctors to reach a diagnosis of paranoid schizophrenia. Following this, the patient developed anxiety and odd movement disorders similar to catatonia (a state of stupor). She was then readmitted into the psychiatric hospital and diagnosed as having had a second psychotic episode. After a few days, she appeared disorientated and had disorganized speech patterns. The results of a brain magnetic resonance imaging (MRI) scan showed no abnormalities although electroencephalogram testing showed signs of encephalopathy, suggesting that the patient may have a brain injury. This led to a new diagnosis of psychosis/schizophrenia with status epilepticus and a possible somatic syndrome. This means that the patient's condition may have resulted from physical symptoms.

Interestingly, laboratory testing showed the presence of antibodies in the serum against the N-methyl-D-aspartate (NMDA) receptor (a type of excitatory glutamate receptor) in the serum and cerebrospinal fluid of the patient. Based on this, a final diagnosis of anti-NMDA receptor encephalitis (a rare disease) was made and the doctors identified a teratoma of the ovary, which was surgically removed. After treatment with anti-inflammatory drugs the patient slowed a slow improvement, although symptoms such as the speech impediments, loss of concentration and altered personality disorder persisted for several months. One year after the final diagnosis, all symptoms had disappeared, and the patient was able to restart her college education.

This case described a patient with a syndrome in which encephalitis is caused by an auto-immune response against an important neurotransmitter receptor. Although this form of encephalitis is considered to be a neurological condition, the clinical presentation can begin with psychiatric symptoms such as anxiety and hallucinations. The lesson here is that biomarker analyses using serological testing for this antibody could have picked up this condition from the start. This would have helped to minimize the duration of illness and reduced the time to recovery.

Case 4

A 68-year-old male admitted to emergency with acute seizures and right hemiparesis, associated with hypoglycaemia. He was referred for a psychiatric evaluation since he had been refusing food and this was thought to be the primary cause of the symptoms. For the previous 8 months, he had been presenting with the delusion that he had neck cancer obstructing his throat with clogged even though these had been ruled out by clinical examination. This delusion led to constant food avoidance and fasting and he had frequent episodes of hypoglycaemia. His psychiatric history included only one episode of depression which had been treated successfully 8 years before, although he displayed behavioural changes over the last 2 years which included hiding food and refusing meals. This was associated with physical effects including sweating, slurred speech and tremulousness which were all relieved by forced eating. He also displayed the clinical manifestations associated with some psychiatric conditions, like self-neglect, lethargy, isolation, confusion, disorganized speech, unresponsiveness and poor performance of normal daily tasks. Because of this, he was admitted to the acute psychiatric ward with a diagnosis of depressive episode with hypochondriac features. He was then treated with the antidepressant venlafaxine and the antipsychotics risperidone and alprazolam. This led to improvements in the eating behaviour problems as well as certain psychiatric symptom scores [Brief Psychiatric Rating Scale (BPRS) and Hamilton Rating Scale for Depression (HAM-D)]. However, hypoglycaemia still occurred even with the normalized feeding behaviour. For this reason, further clinical tests were performed which showed that the patient had very high circulating levels of insulin associated with the low glucose levels. Subsequently, the patient underwent an abdominal computerized tomography (CAT) and transendoscopic ultrasound scans and a tumour was identified and removed. Histopathological testing confirmed this to be an insulinoma. When tested 6 months later, the patient showed no hypoglycaemia and improved psychiatric symptom scores, associated with improved personal and social functioning. Since the patient stilled showed some mild depressive symptoms, the antidepressant was prescribed for 1 more year.

This is another illustration of a case of a suspected psychiatric condition which was actually caused by a somatic condition—in this case the culprit was a tumour called an insulinoma. These tumours produce high levels of insulin, the main hormone responsible for reducing blood sugar levels. Since glucose is a major energy fuel source for virtually all cells in the body (including the neuronal cells in the brain), it is not surprising that the brain would be operating in an energy deprived state and cause the psychiatric symptoms. Again, the lesson to be learned here is that biomarker analyses using blood testing for insulin and glucose would have led to further testing, such as the scans performed above, followed by surgery, which would have helped to minimize the duration of this illness.

The most important point to keep in mind here is that an early and accurate diagnosis can lead to better outcome for the patients (Fig. 2.2).

Current Diagnostic Practices in Psychiatry

Today, rational medicine requires the existence of a valid method to group or classify similar patients using a diagnostic system. This is basically so the right form of treatment can be prescribed for the best possible patient response and outcome. At present, the major psychiatric diagnostic system used in Europe is the Diagnostic and Statistical Manual (DSM) system, with the 2013 DSM-5 as the most recent edition (the method used more commonly in the USA, the International Classification of Diseases Version 10, is discussed later). Even though the influential DSM-III was introduced in 1980, the fundamental concepts of the DSM approach date back to the late nineteenth and early twentieth centuries, when Kraepelin made his observations on the difference between dementia praecox and mania (see previous chapter). Dementia praecox eventually developed as the schizophrenia concept, whereas mania formed the basis for the broad group of mood disorders, which include major depressive disorder and bipolar disorder.

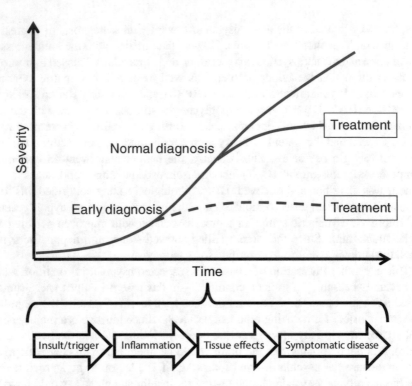

Fig. 2.2 Time-based diagram showing that early diagnosis combined with effective treatment can lead to better outcome for the patients

Most researchers and clinicians are now aware of the fact that the DSM system does not take into account the underlying biological causes of the disorders described in the various diagnostic categories. Instead, it defines these categories using cut-off criteria based on the presence or absence of symptoms. Therefore, these categories within the psychiatric disorders are distinct from other major medical diagnostic concepts, which are mostly associated with underlying biological alterations. Nevertheless, the DSM concepts are not arbitrary and have been chosen based on their clinical validity over approximately 100 years of psychiatric thinking and practice. Indeed, the field of psychiatric research as it is today would not exist without this rigorous system.

It is clear that diagnosis is the principal rate-limiting step in psychiatry and associated research and if clinicians use less than ideal diagnostic categories, the results of these studies will either be absent or clouded by over or under inclusion of patients or controls. Therefore, some or all results derived from such studies might be misleading. The main disadvantage of the DSM diagnostic categories is that they are arbitrary and do not necessarily represent a true medical diagnosis. However, adequate training of clinicians, especially in the use of standardized interviews, can lead to acceptable inter-rater validity using the DSM criteria. Nevertheless, the validity of DSM constructs is limited with respect to providing information on the affected biological pathways as well as delimitation from other disorders. This is because the DSM categories are heterogeneous as they incorporate many combinations of symptoms arranged into each category. For example, schizophrenia can be comprised of 23 different combinations of symptoms and other observations. Another likely problem is that some researchers argue that there is a continuum between the core psychiatric symptoms, syndromes and normal functioning with no actual discrete boundaries. For example, the distinction between sorrow and depression is not clear.

As an example, recent findings show evidence for overlap between schizophrenia and bipolar disorder. Although DSM makes a distinction between schizophrenia and bipolar disorder, Kraepelin's later work pointed to the similarities in the course of both disorders. Furthermore, there is now evidence that autism spectrum disorders and schizophrenia both show brain connectivity deficits and similar genetic variations making these difficult to distinguish in some cases, especially if the subject is an adolescent. Such findings help us to understand that most psychiatric patients do not meet the criteria of one particular disorder as currently defined but instead they can show signs and symptoms of several of these diseases. Figure 2.3 shows this dilemma in a schematic bubble diagram.

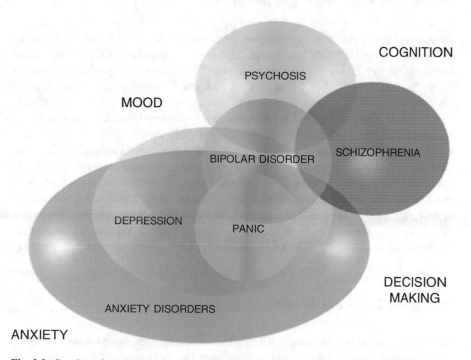

Fig. 2.3 Overlap of psychiatric disorders with respect to effects on various brain functions

How Might Biomarkers Be Used in Psychiatry?

As an Aid to Diagnosis

A logical first step in biomarker discovery would be to investigate an association between biomarkers and a specific diagnostic category. This would require identifying biomarkers which can separate one DSM diagnostic category from another as well as separating patients from so-called *"healthy" control subjects*. However, there are many roadblocks in the way of attempting such an approach. As outlined in the section above, the DSM categories are arbitrary to a certain extent and most likely do not reflect an underlying pathology or biological pathway. This problem is compounded by the fact that most biomarker studies in psychiatry result in considerable overlap across the groups being compared. In other words, they do not tend to make a good distinction across the groups. It would be a breakthrough if a biomarker or set of biomarkers were shown to be specifically related to one or more of the DSM diagnostic categories. However, this would be trivial from a clinical diagnostic point of view, as the category would already be defined by the DSM criteria.

Identification of Diagnostic Subgroups

Another approach is to attempt to bypass the heterogeneity in the DSM categories by going back to the basics of the system. This assumes that there are a limited set of fundamental psychiatric syndromes which can be identified by specific core symptoms. Examples would be: schizophrenia with severe negative symptoms, bipolar disorder in the mania state (more on this later), severe depression with autonomic dysregulation (formerly called endogenous depression) and classical autism. These syndromes are rare but easy to identify due to their severe nature. The first step would be to identify biomarkers that are associated with these syndromes. The next step would be to use these for testing a large group of patients presenting with less specific psychiatric symptoms and attempt to classify these individuals according to their similarities with the fundamental biomarker patterns. The final step would be to investigate whether or not the tested patients with these biomarker profiles have a prognosis or treatment response similar to those of individuals with the associated fundamental syndrome. This would be useful for prescribing the right medications.

Treatment Response Prediction

Biomarkers could also be used for prediction of response of the patients following treatment with specific psychiatric medications. The main goals of this would be to either predict efficacy (effectiveness of treatment) or whether or not the patient would develop adverse events (side effects), or both. Thus, patients could be divided into subgroups of likely responders and non-responders and at risk or not at risk of developing side effects, respectively. The evidence would be used for making decisions on the medications that patients should receive. Examples of biomarkers which could be used for this could include those derived from imaging techniques, serum assays, genetic profiling, physiological measures, histopathological findings or psychological tests.

There are already some examples of genomic biomarker tests that predict response of patients receiving a drug based on how quickly that drug is metabolized within the patients' bodies. Drugs are normally metabolized as they pass via the circulation through the liver by a family of enzymes called the cytochrome P (CYP)-450s. For example, the selective norepinephrine reuptake inhibitor atomoxetine, approved for treatment of ADHD, is cleared from the body through metabolism by CYP2D6. Therefore subjects who have naturally low levels or poor activity of this enzyme tend to have higher plasma levels of atomoxetine compared to those with normal CYP2D6 activity levels as atomoxetine would not be metabolically cleared as quickly in these individuals.

There are other examples of biomarkers that predict differences in responses based on whether excessive levels of a specific gene or protein are present. For example, the *HER2* gene encodes a cell surface receptor protein that causes growth of breast cancer cells and this gene is over-expressed in approximately one fifth of the women who have breast cancers. This is one of the best examples of personalized medicine in clinical practice today. Patients who have over-expressed *HER2* can be given an antibody with the trade name Herceptin® which binds to and neutralizes the HER2 protein and therefore blocks cancer cell growth. There are also a few examples of personalized medicine in psychiatry. Studies have suggested that patients with a particular nucleotide sequence change (this is called a polymorphism) in the serotonin 2A receptor gene (*HTR2A*) tend to have a positive response after treatment with the selective serotonin reuptake inhibitor (SSRI) citalopram. Although these findings are not as striking as those used in cancer, they give encouragement that searching for biomarkers of psychiatric drug response may be fruitful.

Helping to Redefine the Diagnostic Categories

Previously, I have stated that DSM criteria have limited value in helping to understand the underlying biology of these diseases and should therefore not be used to guide biomarker-based research. So, based on this, one approach of increasing our fundamental awareness of the pathways affected in these disorders would be to abandon the DSM categories altogether, using instead the broad array of symptoms and other factors associated with these diseases. Patients could instead be arranged into broad problem groups that are different from one another in terms of disease course, outcome and presumably the aetiology (the cause of the disease). The next step would be to test this problem group based approach in a mixed group of patients to determine if these patients could be segregated into distinct groups based on biomarkers. The mixed group could be comprised of current DSM diagnoses including psychotic disorders, bipolar disorder, depressive disorder and various personality disorders. In this way, patients belonging to the same biomarker group will have the same biological profile (at least for those biomarkers investigated) and may exhibit the specific signs and symptoms of a traditional diagnostic category. However, it is more likely that these clusters will consist of mixture of patients exhibiting signs and symptoms of a variety of DSM-defined disorders. Another aspect which could be tested is whether or not the patients of a particular biomarker group have any common clinical characteristics, such as similar prognoses or responses to treatment. An initial form of this approach has been advocated by Van Praag, who defined some basic symptom groups.

Obviously, there are a number of assumptions associated with this approach. The most important of these is that circulating biomarker differences occur and that these differences reflect underlying changes that are common in a particular subgroup of patients. In support of this possibility, one recent study found that acutely ill schizophrenia patients could be classified into two groups which differed in their serum biomarker profiles compared to the profiles in normal control subjects. One group had more alterations in the levels of circulating hormones and growth factors like prolactin, testosterone and insulin, and the other group had changes mainly in immune or inflammatory factors, such as the interleukins 1RA, 8, 16 and 18. This finding may be important as other reports have suggested that some but not all first onset schizophrenia patients have changes in insulin and other molecules related to insulin action. Likewise, other studies have found that approximately half of schizophrenia patients have changes in molecules associated with the inflammatory response.

Helping to Identify Staging of Psychiatric Diseases

Recent years have focussed more on the dynamic nature of psychiatric diseases as occurs in other disorders. One example of this is the different stages of malignancies in various types of cancer. This is especially true if the origins of a particular psychiatric disorder are neurodevelopmental, since this implies an illness trajectory. For example, four stages of schizophrenia have been hypothesized and these are: (1) risk; (2) prodromal symptoms; (3) psychosis; and (4) chronic disability. Risk is the stage before detectable deficits occur and the prodromal phase of schizophrenia is now known to be a valid second stage which occurs before the onset of full blown psychosis. The prodrome stage is identified based on the presence of symptoms such as disturbed thoughts, social isolation and impaired functioning. Some of these features are common during adolescence and the problem of distinguishing a high risk for psychosis from more common adolescent angst has always been a major challenge. At present, diagnosis of schizophrenia is mainly based on the symptoms and signs of psychosis. However, this is likely to occur not at the beginning but during the course of illness, after a time when some neuronal changes may have already occurred. Chronic disability is the stage of suffering and associated difficulties in patients who have had the disease for several years. The incorporation of biomarker and new cognitive tests, as well as the identification of subtle clinical features, may enable the detection at the earlier stages of risk, such as in the prodrome stage, as well as identification of specific illness trajectories.

What Kinds of Biomarkers Have Been Identified?

Over the last decade, converging results from *post-mortem* research, neuroimaging, genetic association studies and measurements of peripheral blood biomarkers have suggested the presence of several biological themes within the broad context of the schizophrenia syndrome and other psychiatric diseases such as major depressive disorder and bipolar disorder. One of the most recurring themes has been the identification of biomarkers associated with altered glucose metabolism and insulin signalling, growth factor pathways and immunological alterations. In the case of schizophrenia there is also abundant evidence for alterations in dopaminergic- and glutamatergic receptor signalling pathways. In depression these alterations are less clear, but there is long-standing evidence for aberrant HPA-axis signalling (more on this later).

Decreased circulating levels of a protein called brain-derived neurotrophic factor (BDNF) have been identified repeatedly in schizophrenia. Interestingly, this may be dependent on the clinical phase of the disorder as the levels of this growth factor appear to be decreased during acute psychosis and restored to normal levels after remission. Increased levels of inflammation–related molecules, such as interleukin 1, interleukin 6 and tumour necrosis factor, have also been found. But none of this is new. The suggestion of impairments in energy metabolism in psychiatric disorders such as schizophrenia was published almost 100 years ago. Because most of these studies involved taking samples from patients following treatment, hypotheses have emerged and are now established that antipsychotic drugs had a negative impact on glucose metabolism and the insulin response. These days it is well known that antipsychotic drugs such as clozapine and olanzapine can lead to increased body weight, diabetes and hyperlipidemia when given to patients. However, studies over the last 10 years have shown that schizophrenia patients can have insulin resistance, independent of antipsychotic treatment.

In major depression, there have also been numerous reports of alterations in the HPA axis and other hormonal signalling pathways. Evidence for hyperactivity of the HPA-axis was shown by higher levels of the stress hormones CRF, ACTH and cortisol. Also in major depression, convincing evidence of inflammation has been found with increased levels of interleukin 1 and interleukin 6. More details on these biomarker findings are presented in chapters on the major psychiatric disorders schizophrenia, major depression, bipolar disorder, anxiety disorders and autism spectrum disorders, as well as the neurodegenerative disorders Alzheimer's disease and Parkinson's disease. All of these chapters illustrate how we can study these "brain" disorders by investigating the blood.

Future Prospects

This chapter describes the clinical utility of biomarkers for major psychiatric disorders such as schizophrenia, major depressive disorder and bipolar disorder with a focus on the use of a blood test to improve diagnosis and patient outcomes. At present, there are a limited number of clinically valid biomarkers available for this purpose. This may be due to the fact that these are still linked to old diagnostic concepts which have been in use for decades to classify these diseases. It is likely that efforts along these lines will be challenging, due to the heterogeneity inherent in these categories. Moreover, identifying a biomarker for a syndrome that has already been identified based on clinical phenomenology is not useful from the clinical point of view. We need to go beyond this. Innovative approaches are needed such as identification of biomarkers that can be measured in at-risk individuals with prodromal symptoms. It is hoped that tests can be constructed from such biomarkers and that these can be used to assess possible development of the patients towards more severe states, and thus indicate the optimal intervention for that stage. This approach is aimed at disease-profiling and clinical staging and would therefore be based more on readily observable clinical characteristics. Next, it would be important to use broader categories of related patients, and to deconstruct the traditional diagnoses of these patients using molecular biomarker profiles in addition to, or in place of, the symptom based approach used today. Finally, another use of biomarker tests would be for predicting an optimal treatment response, an approach which has already found some success in cancer studies.

Chapter 3
The Importance of Biomarkers: The Required Tools of the Trade

The use of biomarkers for studying disease and developing novel treatment strategies is an emerging approach in many areas of medicine. This new paradigm is aimed at using biomarkers to enable diagnosis and treatment of diseases at the earliest possible time point for the maximum benefit to the patient. There are now many sophisticated biomarker-related technologies in use which are undergoing constant improvement. These include technologies which can identify biomarkers at level of the genes, messenger RNA, proteins and metabolites in bio-samples taken directly from the patients and healthy controls for comparison. Other techniques are emerging which can analyze the function of intact cells, which is one step closer to integrating whole body analysis. This chapter provides basic knowledge of the principles behind these technologies and their application in the laboratory and the clinic. The most important discoveries by these profiling platforms in psychiatric research are also highlighted with an emphasis on how such methods may soon be applied in the clinic to improve the lives of patients.

Why Do We Need Biomarkers in Psychiatry?

Attempts to identify molecular biomarkers for psychiatric disorders have been ongoing for many years. It is thought that biomarkers could be used as a standardized test to facilitate the diagnosis as well as the treatment and monitoring of patients. As described in the previous chapter, these disorders are diagnosed currently by clinicians and psychiatrists based on subjective interviews. However, the currently used DSM and International Classification of Disease 10 (ICD-10) diagnostic classification systems for psychiatric disorders are known to have shortcomings. It is now thought that biomarkers which reflect the underlying disease pathophysiology will lead to improvements in the diagnostic process and pave the way for earlier and more effective treatment of patients.

© Springer International Publishing AG 2017
P.C. Guest, *Biomarkers and Mental Illness*,
DOI 10.1007/978-3-319-46088-8_3

Table 3.1 Classes of biomarkers and associated requirements

Biomarker class	Requirement
Exploratory biomarkers	Scientific evidence for proof of concept
Probable valid biomarkers	Measureable using test systems with strict performance characteristics. Evidence to explain relevance of the test results
Known valid biomarkers	Replication of results in different laboratories and different sites

The development of biomarkers for use in diagnostics and clinical trials is progressing, although this seems to be at tortoise pace. This has resulted in movements towards new standard operating procedures to overcome the current difficulties and to meet the strict demands of the regulatory authorities. This is important as the regulatory agencies have the final word on whether a drug, procedure or biomarker can be implemented in the clinic. However, the FDA now considers that the inclusion of biomarkers into the drug discovery pipeline is an important next phase for the pharmaceutical industry to adopt in order to keep delivering drugs and to be competitive in the current economic climate. In fact, programmes are now underway to modernize and standardize methods to ensure delivery of more effective and safer drugs. As far as biomarkers are concerned, it means that these must be "known validated biomarkers" before they can be used in a clinical trial (Table 3.1). The main requirements for this are: (1) there is scientific evidence that the biomarker is involved in the disease; (2) it is measureable using analyses with strict performance guidelines and the test findings make sense with regards to the disease process; and (3) the results can be replicated in repeat studies at different laboratories. The latter point is critical as this is where most promising biomarkers have fallen by the wayside.

Identification of biomarkers will not be easy to achieve in the case of psychiatric disorders. As discussed earlier, there is a wide spectrum of these conditions and they are notorious for their heterogeneity and overlap in symptoms and how these symptoms are manifested in different individuals. However, new biomarker profiling platforms have emerged which allow the simultaneous measurement of hundreds or even thousands of molecules. This has helped to increase accuracy of the findings, reduce the amount of sample required and lower the overall costs in comparison to techniques which measure one biomarker at a time. Furthermore, recent efforts have seen movements towards miniaturization of biomarker assays on handheld devices which can produce results during the time of a standard visit to the doctor's office. These developments are discussed in the final chapter of this book.

Proteomics

Most researchers consider the study of proteins (proteomics) and metabolites (metabolomics) to be the most informative reflection of biological function, considering that these molecules actually carry out or respond to most processes of the body. As changes in physiological states are dynamic in nature, they are likely to introduce alterations in numerous proteins that converge on similar pathways. Therefore, it would be most useful to apply proteomics in such studies as these methods target tens to hundreds of proteins simultaneously, which should lead to a more comprehensive understanding of the affected pathways. This is like the difference between fishing using specific bait and trawling using a wide net. The first approach catches one type of fish, whereas the second can pull in many different varieties (although this came sometimes be an old boot). With this objective in mind, numerous proteomic studies have already been applied in the study of different brain regions, peripheral body tissues and fluids, which have all been implicated in psychiatric disorders as part of the whole body concept. The approach of profiling bio-fluids like blood serum and plasma is most likely to lead to a biomarker signature with diagnostic and prognostic value, as these are easily accessible through relatively a relatively painless procedure (taking a small amount of blood from a vein) for use in clinical studies. Ultimately, proteomic signatures are expected to be useful for enhancing our knowledge of disease mechanisms and drug actions as well as for the development of new biomarker tests for improved diagnosis, prediction of drug response and for monitoring drug efficacy and side effects.

Most of the studies used in the analysis of brain tissues have led to identification of proteomic abnormalities in energy metabolism, oxidative stress, synaptic transmission or cell maintenance and structure. However, it is still not known whether the effects seen on these pathways represent true disease modifications due to the fact that many of the findings have still not been validated. In addition, the low sample numbers associated with most post-mortem brain studies have resulted in a lack of statistical robustness of the findings. Nevertheless, most of the findings from these different investigations show significant convergence on the above mentioned pathways.

There are other limitations which makes interpretation of the findings from brain proteomics studies difficult. First, the use of post-mortem material is not ideal as it cannot be collected under standard conditions and it therefore varies in such parameters as differences in the time after death that the material is collected and stored (the post-mortem interval), time and method of storage as well as other variables. Furthermore, virtually all psychiatric patients are likely to have received various medications and suffered co-morbidities (other diseases) throughout their lives up to the time of death. Thus, the studies are more likely to uncover the effects of these co-morbidities or the administered drugs than the true disease signature. Also, as mentioned above, studies which have carried out technical validation of the findings using follow up studies have been scarce due to the low availability of high quality post-mortem brain tissues. Many researchers have now come to realize that the application of multiple platforms in combination will not only provide a deeper insight into the affected protein pathways but this will also enable cross-validation of the findings and/or investigation of the abnormalities in a systems-based way.

The next section discusses the major platforms in use for identifying and testing for biomarkers for psychiatric disorders in a clinical environment.

What Are the Main Biomarker Technologies?

Proteomics technologies have seen increasing use at around the same time the Human Genome Project was nearing completion. In 1999, the first proteomics study of cancer was published. As of October 2016, searching PubMed, the primary database for research papers, using the terms "cancer" and "proteomics", returned approximately 13,700 articles, showing the increasing popularity of using proteomic analyses to study these conditions. The first proteomics paper in psychiatric disease was published in the year 2000 and a current PubMed search for the appropriate terms returned only approximately 780 relevant articles. This leads to the question of why proteomic studies of psychiatric disorders have lagged so far behind. One reason for this could be that it is more difficult. Biomarker studies on psychiatric disorders are stymied by the difficulties in accessing the relevant biological materials, since the main manifestations appear to occur in the brain. However as described in this book, there has been a recent movement towards identification of correlating biomarkers in the periphery which should help to narrow the deficit.

Several studies have emerged recently involving investigation of large numbers of serum or plasma molecules at the same time. This is the approach of using multi-protein arrays or mass-spectroscopy profiling platforms. These techniques enable the simultaneous detection of hundreds of proteins, some of which may be found to be altered in disease and can therefore be used as potential biomarkers or as clusters of biomarkers that correlate with disease. Two exciting studies were published almost simultaneously using multiplex protein profiling and these found specific differences between schizophrenia patients and healthy controls.

Mulitplex Immunoassay

Blood serum and plasma samples contain many molecules such as hormones, growth factors and cytokines which require high sensitivity of detection. One of the best methods to achieve this is multiplex immunoassay. However, in order to understand this, the method of standard immunoassay (singleplex) will be described as this illustrates the basic principle. Antibodies against a specific protein are linked into defined wells in a microtitre plate (in the example shown, the antibodies will bind the vital hormone insulin). Then the sample is added and the specific protein is captured by the antibody. Next a fluorescently-labelled secondary antibody against the same protein is added which also binds to the protein, creating a "sandwich" configuration. Each well is then analyzed in a plate reader to determine the amount of bound labelled antibody, which is proportional to the amount of that specific protein in the sample (Fig. 3.1).

The same concept is employed for the individual assays in the multiplex immunoassay technique (Fig. 3.2). This assay is conducted as follows: (1) micro-beads are loaded with different ratios of red and infrared dyes to give a unique fluorescent

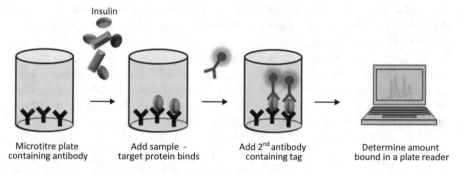

Fig. 3.1 Basic format of "sandwich" immunoassay

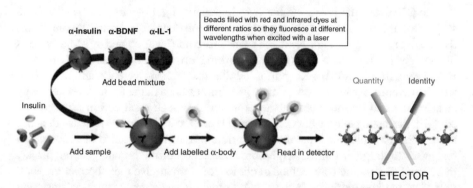

Fig. 3.2 Multiplex immunoassay. Samples are added to dye-coded microsphere-antibody conjugates that target specific proteins. After incubation with a second antibody containing a fluorescent label, the mixtures are passed through a flow cytometry instrument which uses lasers for identification of the antibody-microsphere conjugates and quantitation of the bound analytes

signature; (2) a capture antibody is attached to the fluorescent bead surface such that each specific antibody is attached to a bead with a specific signature (again, an insulin antibody is shown in the example); (3) the target molecule (insulin) binds to the insulin antibody-bead conjugate; (4) a fluorescently-labelled detection antibody is added which binds to target molecule in the sandwich format; and (5) the beads are streamed though a reader and analyzed by two lasers for identification and quantification of the analyte present. The lasers identify which analytes are present by measuring the unique fluorescent signature of each bead, and determine the quantity of analyte present by measuring the amount of the fluorescent tag associated with each bead (proportional to the analyte concentration). In the example shown, it would be possible to determine the concentrations of insulin, brain-derived neurotrophic factor (BDNF) and interleukin 1 (IL1), simultaneously as each antibody-bead conjugate would have a distinct signature as it passes through the reader.

The final step in the process is calculation of the experimentally determined molecular levels of each analyte measured in the samples. As with most immunoassay methods, this is achieved by plotting the readings for each sample along standard curves of known analyte concentrations to derive that of the target molecule. The corresponding concentration determined for that sample can then be adjusted by multiplying by the appropriate dilution factor to calculate the absolute concentration of the molecule of interest. Thus far, this method has been used in the study of multiple psychiatric disorders including schizophrenia, bipolar disorder, major depressive disorder and autism spectrum conditions. This is discussed in more detail in the appropriate chapters describing these diseases.

Two-Dimensional Gel Electrophoresis

Patrick O'Farrell first presented the two-dimensional gel electrophoresis (2DE) in 1975 although he probably did not expect that this technique would revolutionize the discovery of biomarkers. Next, the term "proteome" was coined by Marc Wilkins in 1994 while he was a PhD student working on the concept of the expressed **PROTE**ins of a gen**OME**. Keep in mind that researchers were already using proteomic methods before this time—they just didn't have a name for it. They key thing to keep in mind to understand about proteomics is that, each cell in an organism contains the same genome but only some of these are expressed as proteins depending on, for example, the type of cell, the time of day, the presence of disease, and responses to drugs or toxins. Just before completion of the human genome project in 2001, 2DE remained the method of choice for comparative global proteome analyses. The method works as follows: (1) proteins in the samples are first separated according to their isoelectric point (pI) (the point at which no net charge occurs on the protein) using isoelectrofocusing (IEF), and (2) according to apparent molecular weight (MW) using polyacrylamide gel electrophoresis. The proteins in the gels can be visualized with any number of stains (such as Coomassie Blue or Sypro Ruby) and then quantitated using an imaging software (Fig. 3.3).

   2DE techniques allow the study of intact proteins but there are some problems of using this approach for analyses of blood serum or plasma samples. This occurs mainly due to the fact that blood contains a massive concentration range of proteins spanning more than ten orders of magnitude. This means that very abundant proteins such as albumin would appear as large blobs on the gels and obscure less abundant proteins such as the cytokines.

Mass Spectrometry

Just before the end of the human genome project, the shotgun mass spectrometry revolution began as a new and more sensitive and higher throughput proteomic approach for biomarker identification. These methods are called 'shotgun' as the protein mixtures are digested with enzymes to produce smaller peptides which are

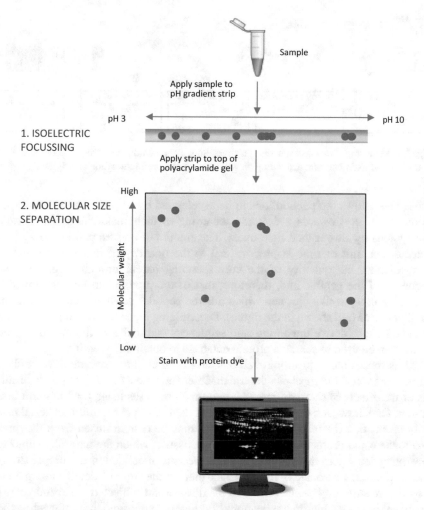

Sample

Apply sample to
pH gradient strip

pH 3 ← → pH 10

1. ISOELECTRIC FOCUSSING

Apply strip to top of
polyacrylamide gel

High

2. MOLECULAR SIZE SEPARATION

Molecular weight

Low

Stain with protein dye

Fig. 3.3 Two-dimensional gel electrophoresis. Proteins are extracted from a tissue or other sample and then separated by electrophoresis in two dimensions. The first dimension is isoelectric focussing, during which proteins are separated according to their isoelectric points (the state of zero net charge). The second dimension is detergent-based gel electrophoresis, during which the proteins are separated according to apparent molecular weight. The resulting protein spots can be subjected to image analysis and quantitated

actually what is analyzed. This is done since proteins are too large and complex in their structure to be analyzed directly. In the next phase, the samples are separated in a high performance liquid chromatography column, basically so that they do not all enter the mass spectrometer at the same time (just picture the opening of a store in the after Christmas sales and you will know what I mean). As the peptides leave the column, they are analyzed by electrospray, which is essentially the application of an electric charge to evaporate all fluids, leaving the peptides in a charged plasma

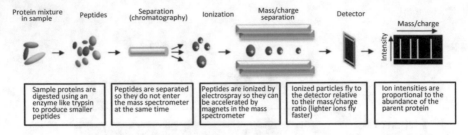

Fig. 3.4 Main stages of mass spectrometry profiling. This method can screen hundreds to thousands of proteins in one run, depending on whether or not a prefractionation step is used

state. This allows the peptide ions to be accelerated by magnets in the mass spectrometer and fly towards a detector according to their mass/charge (m/z) ratios (heavy ions fly slower than light ones). The amount of a given peptide striking the detector per unit of time is proportional to the quantity of that peptide and, by extrapolation, the quantity of the corresponding parent protein (Fig. 3.4). The sequence of the peptide and, therefore, that of the protein can be determined by streaming in a gas like nitrogen which hits the peptides causing them to break into smaller pieces (not shown in the figure). Determining the masses of these pieces can then be used to derive the amino acid sequences that make up the peptides which could then be used to search a protein database to obtain the identity.

Mass spectrometry profiling methods have been used to generate novel molecular information about most psychiatric disorders and most of this has involved studies of the effects of these diseases on neuronal processes using post-mortem brain tissues. One drawback of such studies is that the use of post-mortem material naturally means that the results of these studies may be compromised from the outset due to the usual problems with post-mortem tissues, which are naturally subjected to varying delay periods following death to the time of collection and proper storage and to problems associated with the fact that by the time of death most patients would have suffered chronic effects of the disease and subjected to several years of treatment with psychiatric medications. Taking these confounding effects together, this would make it exceedingly difficult to study the underlying pathways affected in the disease as the true effects may be obscured.

¹H-Nuclear Magnetic Resonance (NMR) Spectroscopy

Although mass spectrometry can be used in analysis of small molecules such as metabolites, ¹H-NMR is mainly used for this purpose as it does not require separation or pre-fractionation of the molecules. One major advantage of this approach is the simplicity of the sample preparation step and the overall analytical reproducibility. ¹H-NMR spectroscopy can give information about the structural properties of molecules and is therefore suited for identification purposes. It works by tracking the behaviour of protons when they are introduced into a magnetic field of very high

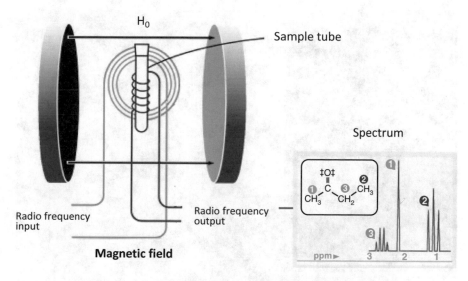

Fig. 3.5 Main principles of ^{1}H-NMR spectroscopy for determining molecular structure

intensity. In this way, the nuclei of the protons in molecules line up directionally with the field, similar to the way in which a compass needle lines up with the Earth's magnetic field (although the magnetic field in an ^{1}H-NMR machine is around 10,000 times stronger in magnitude than the field surrounding the Earth. The ^{1}H-NMR procedure is initiated by applying radio frequency pulses to the sample. This causes the nuclei to change their rotation away from the equilibrium position so that they can rotate around the magnetic field's axis. The frequency of this rotation is related to the chemical and physical environment of the atom within the molecule. Then by using different combinations of radio pulses it is possible to determine how each atom interacts with other atoms in the same molecule, yielding the structure and therefore the identity of the molecule (Fig. 3.5). ^{1}H-NMR spectroscopy can be used to monitor relative changes in the levels of key analytes relevant to the study of psychiatric disorders such as amino acids, neurotransmitters and neurotransmitter precursors or metabolites.

Imaging Techniques

Another way of studying disease signatures in live patients can be achieved using imaging approaches such as positron emission tomography (PET), which enables direct analysis of brain function. PET imaging first requires the design of a molecule called a ligand which binds tightly and specifically to a single target molecule. The ligand is labelled with a positron emitting radioisotope with a short half life ($t_{1/2}$), such as ^{18}F ($t_{1/2} \sim 110$ min). Following intravenous administration of the radiolabelled ligand, emitted positrons strike nearby electrons resulting in the production of pairs of photons that travel in opposite directions to each other. These photons are

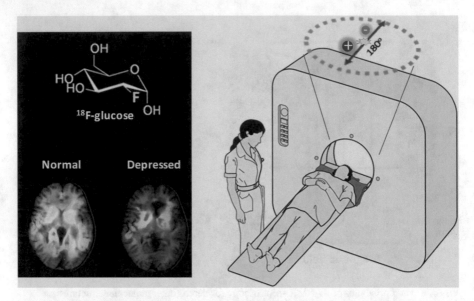

Fig. 3.6 Main stages of how a PET scanner works. Following intravenous administration of the radiolabelled glucose (F-glucose, also known as fluorodeoxyglucose—FDG), emitted positrons strike nearby electrons resulting in the production of pairs of photons that travel in opposite directions (180°) to each other. These photons are then detected by a gamma radiation detector allowing the spatial distribution of the glucose to be reconstructed. This allows the study of changes in regional brain as found in some psychiatric disorders such as depression. The insert shows the structure of ^{18}F-glucose (*top*) and the lower metabolic activity (*yellow colour*) in the prefrontal cortex of a patient with major depression compared to a "normal" subject (*bottom*)

then detected by a gamma radiation detector which surrounds the individual being tested, allowing the spatial distribution of the ligand to be reconstructed. PET can be used in the clinic to study regional brain metabolism changes as often seen in psychiatric disorders using a radiolabelled analogue of glucose called ^{18}F-fluorodeoxyglucose (FDG) (Fig. 3.6). The hypothesis is that these changes in metabolism are indicative, and may even be causative, of changes in neuronal activity. One of the main discoveries in major depression is the finding that many patients with this disorder have increased metabolism in the amygdala and decreased metabolism in the prefrontal cortex. This is thought to be an indication of altered connectivity between these two brain regions. The effects of this are discussed in further detail in the chapter on major depression.

Future Prospects

This chapter has described the issues and current technologies associated with the discovery and development of blood-based biomarkers for psychiatric disorders. The current diagnostic process and ways of studying the biological basis of these disorders are in need of an overhaul. One way of achieving this is through the development of new proteomic-based approaches. Following the completion of the Human Genome Project (HGP), the Human Proteomics Organisation (HUPO) emerged as the next stage, with the aim of deciphering the complexities of modern medicine through the identification of biomarkers. As stated by the Nobel Laureate Lee Hartwell who was a guest speaker at the 2004 Human Proteomics Organisation (HUPO) meeting in Beijing, "biomarkers for early diagnosis will revolutionize the pharmaceutical industry allowing diseases to be treated at an earlier stage—increasing survival rate" (I was fortunate enough to attend this same meeting and—I have to admit—I was hooked). Nevertheless, there is still a reluctance to accept the idea that biomarkers will be of any help at all. It is true that only a small handful of the hundreds of promising biomarkers that have been identified and studied have lived up to the initial hype. Most have actually fallen by the wayside and are now gathering cobwebs. There are several reasons for this. The HGP had a clear task of sequencing the 3.2 billion nucleotides comprising the genome but proteomic scientists wonder if their self-administered task is even possible. "Genes were easy", stated Samir Hanash who headed the HUPO from its inception in 2001. However, a complete cataloguing of the entire set of proteins in the human body is a quest that could take hundreds of years, if it is even possible. The human genome is now known to contain 20,000–25,000 genes which code for proteins. But the actual number of proteins is expected to exceed that by at least one order of magnitude (more likely, this is two orders of magnitude). This is due to many factors such as alternative splicing of genes and post-translational modifications of the proteins following biosynthesis by processes such as limited proteolysis, glycosylation and phosphorylation. All of these processes can give rise to multiple different forms of the same gene product. Another factor which makes the process difficult is that proteins do not tend to work alone. In fact, most proteins are known to work together in complexes or signalling pathways. For these reasons, most scientists think that progress on deciphering the proteome will depend on new breakthroughs in technology. As the reader continues on, he or she will discover that this is already happening.

Part II
Psychiatric Diseases

Chapter 4
Schizophrenia and the Mind–Body Connection

This chapter describes one of the most debilitating and severe psychiatric conditions—schizophrenia. This disease affects approximately 1 % of the people on the planet and, as such, it fulfils the criteria of a major pandemic. It is characterized by diverse and disconcerting symptoms in those who suffer from this disease, such as psychosis, hallucinations, delusions, depression, disrupted thinking and cognitive deficits. But the situation may be even more complex than that. Many researchers have started to adopt the viewpoint that schizophrenia may actually be a single term for a diverse set of different psychiatric disorders. In addition, the diagnosis of schizophrenia is not straightforward and there is still not a complete understanding of the underlying molecular causes. This chapter describes the current clinical needs exemplified by the mediocre success rates of diagnosing and treating individuals with schizophrenia. Inaccurate diagnoses and inadequate treatment will not only hurt the patients but it will also have a negative impact on the patients' families, society in general, and on cost management of the healthcare services. This chapter also provides details on recent advances in screening patient blood samples for identification of biomarkers as the first step towards improving patient outcomes. Most importantly, the finding that many schizophrenia patients also show signs of perturbed metabolic and inflammatory systems may offer a unique solution for improved treatment approaches, involving co-administration of anti-diabetic or anti-inflammatory compounds along with the traditional antipsychotic medications to kerb these symptoms. At the same time, such an approach may also help to alleviate some of the symptoms associated with the disease itself.

How Do We Currently Diagnose Schizophrenia?

Schizophrenia is a devastating psychiatric disorder that affects approximately 1 % of the people in the world at some point in their lifetime. The main symptoms can include delusions, hallucinations, disorganized thinking, anhedonia, blunting, low

© Springer International Publishing AG 2017
P.C. Guest, *Biomarkers and Mental Illness*,
DOI 10.1007/978-3-319-46088-8_4

motivation, social retreat and cognitive impairments. Diagnosis of this disorder is currently based on expression of these symptoms and relies on interview-based communications between physicians and patients. Categorization of patients is achieved using the DSM system as described earlier or using the ICD-10. The latter is mainly used in the USA and the former in Europe. However, these manuals only describe and categorize the various symptoms of psychiatric disorders and do not provide physical or molecular correlates of the disease. Furthermore, they do not attempt to explain the underlying physiological pathways which are affected in the disease state. Put simply, they aim to tell us what—but without telling us why. Disease classification can also be confounded by the fact that other psychiatric disorders can exhibit similar or overlapping symptoms. For example, individuals who suffer from anyone of the three major psychiatric conditions of schizophrenia, major depression and bipolar disorder can all show signs of psychosis and depression. It is easy to see that such an overlap in symptoms can contribute to the problem of frequently occurring misdiagnoses.

Given these difficulties, the development of new tests for molecular biomarkers would be an aid to diagnosis of schizophrenia and may even help to stratify patients into subcategories. The highest impact would be gained if these biomarkers could be identified in the blood as this would lend itself more readily to clinical testing. Blood-based biomarkers are more clinically suitable than biopsied material due to ease of accessibility, low invasiveness of the sampling procedures and lower costs. Recent studies have shown that some schizophrenia patients have altered serum concentrations of molecules involved in inflammation or the immune response. Other studies have shown that schizophrenia patients can display perturbations in metabolic or hormonal pathways in both the blood and the brain. In addition, there have been reports of HPA axis disturbances in some schizophrenia patients, which have been linked to abnormal insulin signalling. The fact that such changes in the peripheral regions of the body can be associated with altered functioning of the mind supports the concept of a connection between the brain and body, at least in the pathways which are perturbed in psychiatric diseases.

Can This Be Improved By Incorporation of Biomarkers?

Early onset of schizophrenia can have long-lasting impairments on social functions throughout life which, in turn, will have major implications on health, family life, the workforce and society in general. Schizophrenia is thought to develop through prodromal (precursor) phases in late adolescence or early adulthood and the actual disease onset is marked by display of overt psychotic symptoms. There are high rates of unemployment and hospitalization associated with schizophrenia, along with the need in some cases for lifelong care. All of this leads to high treatment costs which, in turn, cause an inflated financial burden on society that is greater than that of most other illnesses. It is striking that a high proportion of these costs result from the clinical impact of co-morbidities such as coronary heart disease, weight gain and diabetes, which are common side effects of treatment with second-generation antipsychotic drugs. To complicate matters, fewer than half of the patients diagnosed with schizophrenia respond favourably to initial treatment and many come off their medications, leading to more hospitalizations with a greater length of stay and care costs. This is due to that fact that current treatments are not necessarily designed to target the perturbed biological networks in patients. Most current treatments for schizophrenia are directed towards different combinations of dopamine, serotonin and norepinephrine receptor systems and there are no clear guidelines for physicians in the selection of the most appropriate drugs for each patient. This could be an unavoidable consequence of the practice that current diagnoses mainly detect symptoms of psychosis although a range of other symptoms can also occur, which may be associated with a variety of different pathophysiologies. However, it is hoped that biomarkers can be used to improve diagnostic accuracy and stratification of patients by helping to define disease subtypes and even symptom-based categories. This will facilitate more effective treatments and better patient care by helping to place the right patients on the right treatments which target their illnesses more directly.

Schizophrenia Symptoms

Individuals with schizophrenia may hear voices that are not heard by others, they may believe other people are reading their minds or planning to harm them, they often have difficulty maintaining employment or personal care or hygiene and they may not make sense when they speak. Any or all of these behaviours can make them feel withdrawn or agitated.

The symptoms of schizophrenia can be grouped into three main categories termed positive, negative and cognitive:

Positive Symptoms

This can sound a bit paradoxical but "positive" symptoms are psychotic behaviours that are not observed in healthy people. Sometimes these symptoms can be severe and at other times hardly noticeable, depending on whether or not the patient is undergoing treatment with an antipsychotic drug. Perhaps the most well known of the positive symptoms is hallucination. This can be things that a person sees, hears, smells, or feels that are not really occurring, at least at that time. Hearing voices is the most common hallucination. The imagined voices could tell the individual about their own behaviour, tell them to do things or warn them of danger or threats. Delusions are another common positive symptom. These are false ideas or beliefs which can persist even if others have proven that these beliefs are not true, idiotic or logical. As an example, schizophrenia patients may believe that others can control their behaviour or that people on television or the radio are sending special messages just to them. Along the same lines, they may develop paranoid beliefs that others are trying to harm them, such as by poisoning, or spying on them or plotting against them. They can also believe that they are a famous or infamous historical figure and act accordingly. Another positive symptom is thought disorder, which includes unusual or disorganized ways of thinking. These individuals may talk in a convoluted way that is hard for others to follow or understand. Some schizophrenia patients may stop speaking abruptly or they might use meaningless words called "neologisms" to get their points across. Schizophrenia patients may also show repetitive or agitated body movements.

Negative Symptoms

Perhaps more appropriately named, this category is the loss of normal emotions and behaviours. However, negative symptoms can be more difficult to detect and when identified could be mistaken as a sign of depression or another psychiatric condition. One of the most common of the negative symptoms is a flattened affect in which the individual may talk in a dull or monotonous tone. Negative symptoms could also include a displayed lack of pleasure in all aspects of life, or a lack of initiative or low verbosity. Schizophrenia patients with negative symptoms may also need help with ordinary everyday tasks, or they might appear lazy or even neglect basic personal hygiene and other personal needs. In all, the negative symptoms are thought to contribute more to the poor functional outcomes and low quality of life for people with schizophrenia than the positive symptoms. In addition, the care providers of schizophrenia patients with negative symptoms have also reported higher levels of burden. This may also result from the evidence that negative symptoms persist longer throughout life and they are notoriously more difficult to treat.

Cognitive Symptoms

These tend to be the hardest to detect. As with the negative symptoms, the cognitive symptoms may also be mistaken as part of another disorder. They are usually only detected patients are tested directly. These symptoms can include an inability to understand information or to use information to make decisions. Cognitive symptoms can also be manifested as difficulties in focussing and attention, or problems with working memory (the ability to act on new information after learning it). As with the other symptom categories, the presence of cognitive symptoms often makes it hard for the affected individuals to lead a normal life and this can cause a significant amount of emotional distress.

Diagnostic Tools

DSM

The six diagnostic criteria for schizophrenia (A-F) have changed in DSM-5, compared with earlier versions of DSM. In criterion A, diagnosis has now expanded to include five symptoms and at least two of these are required to be present for at least 1 month.

These are listed below:

1. **Delusions**
2. **Hallucinations**
3. **Disorganized speech**
4. **Grossly disorganized or catatonic behaviour**
5. **Negative symptoms** (diminished emotional expression or avolition)

In addition, one of the two required symptoms must either be delusions, hallucinations, or disorganized speech. Thus, no single symptom is necessarily characteristic of schizophrenia, and this again illustrates its heterogeneity and the probability that it is actually more than one disease. To fulfil criterion B, the level of functioning should be markedly below the level achieved prior to disease onset (this is not a criterion for schizoaffective disorder). Also, if symptoms of schizophrenia begin in childhood or adolescence, the expected level of function should not be attained. Criterion C requires a 6 month duration that distinguishes schizophrenia from schizophreniform disorder (1–6 month duration) and brief psychotic disorder (duration of 1 day to 6 months). Criterion D distinguishes schizophrenia from schizoaffective disorder by the presence of psychosis and more consistent mood symptoms. Finally, criterion E rules out the possibility that psychosis has occurred due to intake of drugs or presence of a medical condition and criterion F makes the distinction between schizophrenia and an autism-related condition.

Definitions
Schizoaffective disorder—a diagnosis made when the patient has features of both schizophrenia and either bipolar disorder or depression but does not meet criteria for either disease alone.

Schizophreniform disorder—a diagnosis made when symptoms of schizophrenia are consistently present within a 1-month period but not for the full 6 months required for a schizophrenia diagnosis.

Positive and Negative Syndrome Scale (PANSS)

The PANSS is sometimes employed during research for measuring symptom severity of patients with schizophrenia and related disorders and can be used for the studying the effects of antipsychotic therapy. To carry out the test, an interview is conducted that lasts approximately 45 min during which the patient is rated with a score of 1–7 on 30 different symptoms covering the positive, negative and general aspects of the disease, with a rating of 7 being the most severe.

Positive scale (7 items, maximum score = 49)
Delusions (unfounded, unrealistic or idiosyncratic beliefs)
Conceptual disorganization (tangential, illogical, loose or blocked thinking)
Hallucinations (auditory, visual, olfactory or somatic perceptions that are unfounded)
Hyperactivity (accelerated movement, response to stimuli, mood lability or hypervigilance)
Grandiosity (exaggerated self-opinion, convictions of superiority, abilities or moral righteousness)
Suspiciousness/persecution (exaggerated guardedness and distrust)
Hostility (expressions of anger resentment or sarcasm)

Negative scale (7 items, maximum score = 49)
Blunted affect (diminished emotional responsiveness)
Emotional withdrawal (lack of interest in life events)
Poor rapport (lack of interpersonal empathy and reduced communication)
Passive/apathetic social withdrawal (diminished interest and initiative insocial interactions)
Difficulty in abstract thinking (difficulties in classifying or generalizing in problem-solving tasks)
Lack of spontaneity and flow of conversation (poor communication due to apathy or avolition)
Stereotyped thinking (decreased fluidity due to rigid, repetitious or barren thought content)

General scale (16 Items, maximum score = 112)
Somatic concern (physical complaints or beliefs about bodily illness or malfunctions)
Anxiety (nervousness, worry, apprehension or restlessness)
Guilt feelings (sense of remorse or self-blame for real or imagined misdeeds in the past)
Tension (physical manifestations of fear, anxiety, stiffness, tremor, sweating and restlessness)
Mannerisms and posturing (unnatural movements or posture)
Depression (feelings of sadness, discouragement, helplessness and pessimism)
Motor retardation (slowing or lessening of movements and speech, diminished responsiveness)
Uncooperativeness (refusal to comply, including distrust, hostility and rejection of authority)
Unusual thought content (thinking characterized by strange, fantastic or bizarre ideas)
Disorientation (lack of awareness of persons, place and time)
Poor attention (failure in focused alertness seen as poor concentration and distractibility)

Lack of judgment and insight (impaired awareness of psychiatric condition and life situation)

Disturbance of volition (disturbance in control of thoughts, behaviour, movements and speech)

Poor impulse control (poor control of urges, leading to sudden release of tension and emotions)

Preoccupation (focus on internal thoughts and feelings to the detriment of reality orientation)

Active social avoidance (reduced social involvement with unwarranted fear, hostility or distrust)

PANSS total score (30 items, maximum=210)

Brain Effects

There is still no cure for schizophrenia although many pharmaceutical and behavioural approaches have been tested. Current research shows that only around one in five individuals recover significantly to be classified as "cured". The most common treatment is pharmaceutical with the administration of antipsychotic drugs. Neurological studies of the brain have shown widespread abnormalities in the connectivity of different brain regions in the affected individuals. The symptoms appear to result from either under or over release of certain neurotransmitters between brain regions (described in further detail below) (Fig. 4.1).

New imaging technologies have allowed researchers to study the structure of brains from schizophrenia patients using magnetic resonance imaging (MRI) and magnetic resonance spectroscopy (MRS) approaches. The most common result of these studies was the finding of an enlargement of the lateral ventricles, the fluid-filled sacs that surround the brain. There is also evidence that the volume of the

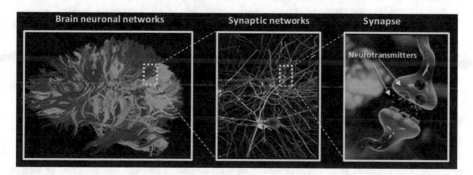

Fig. 4.1 A prevailing theory is that disrupted neural connectivity in schizophrenia results from an imbalance in neurotransmitters. The image on the left was produced by a consortium of more than 100 neuroimaging scientists at ten institutions in a multibillion dollar endeavour called the Human Connectome Project, aimed at providing a detailed map of the neural networks of the human brain

Fig. 4.2 Coloured positron emission tomography (PET) brain scans of a normal person (*top*) and a schizophrenia patient (*bottom*) during a speaking task. The normal brain shows activity in prefrontal and motor areas, with some activity in the parietal area (*red/yellow colours*). The schizophrenia brain shows decreased activity in the prefrontal area, with higher activity in the parietal area and temporal gyrus

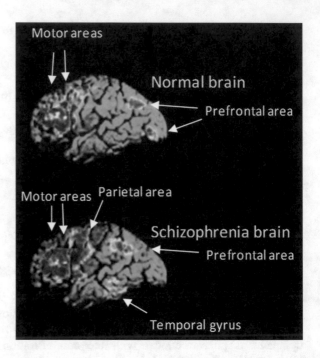

brain is reduced and that the cerebral cortex is smaller in some patients. Accordingly, blood flow was found to be lower in frontal regions of the brain in patients compared with controls. In agreement with these studies, other investigations showed that individuals with schizophrenia have reduced activation in several frontal brain areas while carrying out tasks in which these regions are activated in normal people (Fig. 4.2). There is also evidence showing that the temporal lobe region is smaller in schizophrenia patients and some studies have suggested that the hippocampus and amygdala may also have reduced volumes. Other regions which have been shown to smaller in schizophrenia patients include parts of the limbic system which are involved in control of mood and emotions, and language-related areas. Interestingly, the severity of auditory hallucinations has been found to be associated with the sizes of these language areas. The area of the brain that has been found to be most severely affected is the prefrontal cortex. This part of the brain is associated with memory, and dysfunction here might explain the well-known disordered thought patterns found in schizophrenia. The prefrontal cortex has high concentrations of dopamine pathways and therefore targeting these receptors may be useful for improvement of cognitive symptoms.

Proteomic studies of post-mortem brain tissues in psychiatry have been mainly carried out using the 2DE and shotgun proteomic techniques described in the previous chapter. Several individual proteins as well as entire protein networks have been implicated based on these findings. For example, changes have been identified in proteins involved in nerve cell structure, oxidative stress pathways, metabolism and neurotransmitter production and release. Taken together, the changes in these proteins suggest that there is net effect of diminished synaptic function.

It is still not known if the observed changes arise first in the cascade of events leading to these disorders or if they are a consequence of the cellular and structural alterations which have ensued as a result of the schizophrenia during the lives of these patients. The strongest evidence for neurochemical changes have been found in the dopaminergic, glutamatergic and GABA systems. This has led to the hypothesis that a hyper-responsive dopamine system might be the neurochemical correlate of psychotic symptoms.

Another hypothesis of schizophrenia suggests that neurodevelopmental perturbations can precipitate the disease long before the onset of symptoms. One piece of evidence for this comes from the findings that many susceptibility genes for schizophrenia have critical functions during neurodevelopment. For example, the genes *DISC1*, *NRG1* and *DTNBP1* all have roles in cellular migration, proliferation and synaptogenesis, processes which are required in normal brain development. Also, environmental disturbances such as maternal malnutrition, stress and infection during gestation as well as birth complications are thought to have adverse effects on early neurodevelopment, thus increasing the risk of developing schizophrenia. Besides these findings, subtle robust brain structural changes identified during postmortem studies or using brain imaging approaches have indicated abnormal neuronal proliferation, differentiation, migration and organization. Reductions have been identified in regional and overall brain volumes concomitant with enlargement of the open spaces known as ventricles in first-onset schizophrenia patients and these effects may progress over time. Consistently, neuronal disarray and reduction of neuronal size have been found in specific brain areas of schizophrenia patients such as the hippocampus and frontal cortex. Furthermore, synaptic alterations have been implicated in schizophrenia through molecular studies and may be an indication of altered synaptic formation.

Neurotransmitter Theories of Schizophrenia

The current understanding of the neurochemistry of schizophrenia first came from binding studies of key receptors in post-mortem brain samples from schizophrenia patients. This information has now been combined with in vivo PET and other related imaging techniques such as SPECT (single photon emission computed tomography).

The various neurotransmitter theories of schizophrenia are given below:

Dopamine

The dopamine hypothesis was the first theory developed to explain the pathophysiology of schizophrenia. Most of the supporting evidence for this comes from the indirect observation that many antipsychotics work by blocking dopamine neurotransmission and amphetamines which can induce psychosis-like effects by activating dopamine signalling. In addition, post-mortem binding studies and PET imaging analyses have found increased dopamine receptor binding in the brains of schizophrenia patients. Dopamine signalling controls many high level brain functions which are known to be affected in schizophrenia and in other psychiatric disorders. These functions include motivation, reward and cognition, and other processes such as motor control which are also affected in Parkinson's disease and some movement disorders. Although the neurons regulated by dopamine are present in only a small number of brain areas, they are connected through projections to several other parts of the brain. One of these is the substantia nigra which in turn is connected to other brain regions like the striatum, which regulates motor control. The largest group of dopaminergic neurons in the human brain lies within the ventral tegmental area which is connected to the prefrontal cortex, nucleus accumbens, and other regions involved in regulation of reward and motivation. Some neurons in the arcuate and periventricular nuclei of the hypothalamus are connected to the pituitary gland and regulate secretion of the hormone prolactin. It should be noted that this circuit provides a direct link between the brain and periphery since prolactin is secreted into the bloodstream to control many peripheral functions and these, in turn, regulate many brain functions. When dopamine levels are low, the pituitary secretes prolactin into the circulation whereas high dopamine levels inhibit prolactin secretion. Likewise, the neurons of the zona incerta project to the hypothalamus, which regulates release of gonadotropin-releasing hormone, the hormone which controls development of the reproductive systems in males and females during and after puberty.

Dopaminergic pathways also connect the prefrontal cortex to the ventral tegmental area and the nucleus accumbens. This interconnected network controls motivation which converts needs and goals into action. The dorsolateral prefrontal cortex (DLPFC) is the main integrating hub in this system and impairment of this region could result in lack of motivation as seen in depressive disorders. Considering that most antipsychotics lead to reduced dopamine levels, the DLPFC is also a key area of interest in schizophrenia studies.

It should be stressed that the dopamine hypothesis only accounts for one aspect of schizophrenia, namely the positive symptoms like psychosis. Therefore, other neurotransmitter systems are also likely to be involved.

Glutamate

The glutamate hypothesis came about from the identification of reduced levels of this neurotransmitter in the cerebrospinal fluid of schizophrenia patients, although these findings were disputed soon after. It was next proposed that schizophrenia may result from a deficiency of the NMDA subtype of glutamate receptor neurotransmission, based on the findings that PCP and ketamine, which target this receptor as blocking agents (antagonists), can lead to psychotic symptoms in normal healthy subjects. In addition, post-mortem studies have found reduced NMDA receptor subunit expression in key areas of the brain such as the hippocampus and prefrontal cortex. However, glutamate is widely produced throughout the brain as the main excitatory neurotransmitter and disruptions in this system do not necessarily account for the localized neurological disturbances observed in schizophrenia. A more likely scenario is that it plays a role in coordination with other neurotransmitter systems.

GABA

The GABA neurotransmitter system, the main inhibitory neural network in the brain, which dampens down responses of the excitatory pathways, has also been implicated in the pathophysiology of schizophrenia. Post-mortem studies have found reduced GABA expression in the prefrontal cortex region of schizophrenia patients by looking at the messenger RNA levels of glutamic acid decarboxylase, the rate limiting enzyme in the synthesis of GABA. It has also been proposed that up regulation of $GABA_A$ receptors may occur as a compensatory mechanism in response to the reduced GABA neurotransmitter levels.

Acetylcholine

Most schizophrenia sufferers are also tobacco smokers and this may be an act of "self-medication" to compensate for a deficit in nicotinic acetylcholine receptor neurotransmission. In fact, tobacco smoking may help to improve both positive and negative symptom domains in schizophrenia patients. Post-mortem studies have also shown reductions in muscarinic acetylcholine receptors in key regions of the brain which have been implicated in schizophrenia pathology.

Serotonin

The serotonin system has also been implicated in schizophrenia since antagonism of this system using the broader spectrum antipsychotics clozapine and risperidone can have therapeutic benefits. However, direct evidence for serotonergic dysfunction as a potential disease mechanism in schizophrenia is lacking.

The Link Between the Brain and the Periphery

For decades most psychiatrists and scientists have acted on the idea that mental illnesses are caused solely by defects within the brain. However, developments in various areas of science during the last twenty years or so have led to new concepts that involve the whole body in the precipitation and course of these conditions (remember

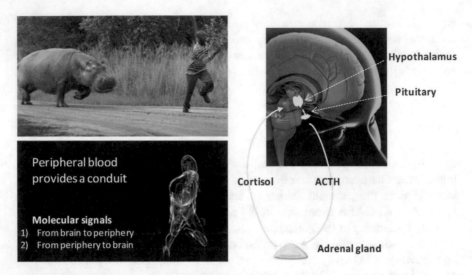

Fig. 4.3 The fight or flight reflex. This is the sequence of molecular events transmitted through the bloodstream which enables a response to sensed danger. Therefore the molecules in the blood can be used as biomarkers of this state

that the ancient Greeks already knew this quite some time ago but we apparently forgot—see Chap. 1). We now know that the brain is holistically integrated in the most fundamental biological functions of the body and this can be reflected by changes in the molecular composition of the blood. In other words, the blood is like a highway network and the molecules are like vehicles which transport messages back and forth between headquarters (the brain) and the outposts (the peripheral tissue). One of the best examples for this is the fight-or-flight reflex seen as changes in the circulating levels of ACTH and cortisol, which provides the molecular and physiological building blocks for the required muscular contractions involved in the reflex (Fig. 4.3). In the example shown, the man makes what is undoubtedly the correct decision not to stay and fight the hippopotamus (don't worry—he got away).

Another example is the sex hormones. Prenatal exposure of both oestrogen and testosterone is known to be involved in early brain development. Interestingly, the ratio of the index finger to ring finger length (2D:4D) has been found to be an indicator of the intrauterine exposure to sex hormone levels. Also, studies have shown that increased testosterone in adolescent males may change dopamine responsivity, one of the main neurotransmitter systems thought to be involved in schizophrenia (see above). A similar study showed that high salivary testosterone levels could be used as an indicator of schizophrenia risk in adolescent males. Another investigation showed that serum testosterone levels significantly predicted performance on memory and processing speed in men with schizophrenia. Also, a positive correlation has been found between progesterone levels and emotional responses in males with schizophrenia as well as in control males.

In line with this, multiplex immunoassay profiling studies of schizophrenia serum or plasma samples have identified changes in hormones and growth factors.

Many of these changes could be explained by the well known dysregulation of the HPA axis in schizophrenia as described above in the fight or flight response. In turn, HPA axis activation can have effects on neurotransmitter systems throughout the brain which can influence mood and behaviour. One study found increased levels of the adrenal hormone cortisol in schizophrenia patients at their first onset and other studies have found that cortisol levels are correlated with the severity of negative symptoms. The impact of this could be far reaching as the HPA axis is only a defined loop of the diffuse neuroendocrine system which includes other peripheral hormone-producing tissues such as the distinct gonadal systems in males and females, the thyroid gland, adipose tissues, the L-type cells of the gut and pancreatic islet cells.

Perhaps the best studied of the peripheral hormone producing tissues are the beta cells of pancreatic islets. These cells are responsible for producing and releasing the glucoregulatory hormone insulin and several other bioactive proteins and peptides (and even neurotransmitters which may act distally or locally). Studies over the last decade have found impaired fasting glucose tolerance, high insulin levels and insulin resistance in first episode patients who had never taken antipsychotics up to that point. Similar effects have been found in drug-free chronic schizophrenia patients. This is likely to result in problems on mood and behaviour since chronically high circulating insulin levels can have deleterious effects on brain function. These effects include causing increased concentrations of pro-inflammatory cytokines, altered phosphorylation of structural proteins, increased β-amyloid plaque deposition (as found in Alzheimer's disease) and perturbed function of neurotransmitter systems and synaptic plasticity.

The production and release of most hormones is regulated by an oscillating feed forward–feedback relationship between the brain and neuroendocrine cells throughout the body (Fig. 4.4). This probably explains why high insulin levels are associated with increased secretion of prolactin and growth hormone from the pituitary,

Fig. 4.4 Diagram showing the neuroendocrine link between the brain and periphery through the circulation via the release of hormones. Note that this diagram only shows the links between the brain, pituitary, pancreatic islet cells and the adrenals. There are many other glands associated with the diffuse neuroendocrine system, such as the thyroid, gonads and intestinal L cells

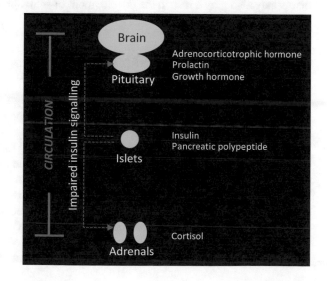

along with altered secretion of chromogranin A and pancreatic polypeptide from other cell types within pancreatic islets and perturbations in the release of the sex hormones estradiol, testosterone and progesterone in first onset schizophrenia patients. Thyroid function has also been linked to effects on insulin signalling and other studies have also found decreased serum levels of thyroxine, tri-iodothyronine and thyroid stimulating hormone in schizophrenia patients.

Most of the proteomic studies carried out using mass spectrometry-based approaches on blood serum or plasma from schizophrenia patients have found changes in proteins involved in functions such as transport of small molecules (like lipids and iron) or the clotting cascade, which leads to changes in wound healing. In addition, multiplex immunoassay profiling studies have found changes in several proteins related to inflammation and immune function. For example, profiling studies found increased levels of inflammation molecules known as cytokines in the cerebrospinal fluid of first-episode schizophrenia patients. This result suggests that the immune system is activated in the brains of some patients and these findings are in line with studies showing that brain development can be affected by changes in the interplay between pro-inflammatory and anti-inflammatory pathways. Another study showed that some serum cytokines such as interleukin 1β, interleukin 6 and transforming growth factor beta (TGF-β) may be biomarkers for acute exacerbations of schizophrenia symptoms and could therefore be used as biomarkers for changes in state of the condition. In contrast, others proteins such as interleukin-12, interferon-γ and tumour necrosis factor-α appeared to be altered in multiple states of schizophrenia and may therefore constitute general schizophrenia trait biomarkers. Large scale multiplex profiling studies have also found changes associated with immune function and the acute phase inflammatory response, including effects on proteins such as alpha 1 antitrypsin, C-reactive protein, factor VII and macrophage migration inhibitory factor. Many of these proteins have also been found to be changed in patients with auto-immune diseases. This is consistent with the idea that autoimmune mechanisms may also play a role in the development of schizophrenia.

Are There Any Novel Treatments on the Horizon?

To recap, some first onset schizophrenia patients have high insulin levels or increased insulin resistance. Interestingly, many patients also show increased levels of this hormone along with other metabolic side effects such as weight gain after treatment with second generation antipsychotics like olanzapine. Indeed, some scientists and clinicians have even postulated these are not really side effects but they may actually be involved in the therapeutic mechanism of action. In other words, one might have to gain weight in order to get better. However, this could still be damaging to aspects of health as found for some overweight and obese individuals who often suffer from devastating conditions such as metabolic syndrome, diabetes or cardiovascular disorders. In addition, the weight gain can be distressing to some patients. This is one reason why many discontinue medication and then suffer a relapse.

However, there may be hope here. Patients who have high levels of insulin either before or after treatment might benefit from the co-administration of drugs which improve insulin receptor signalling, as a means of managing the weight gain. With this in mind, a study tested the use of the insulin sensitizing agents metformin and rosiglitazone to correct the antipsychotic-induced insulin resistance typically associated with this class of drugs. This approach was successful on two counts in that it did curb the weight gain but also without compromising the antipsychotic benefits on mood. For maximum benefit and using a personalized medicine approach (see Chap. 10), future studies testing the effects of this add-on treatment should first aim to select those patients who are most likely to benefit using insulin-related biomarkers for stratification and for monitoring treatment responses or side effects. Stratification is critical as most clinical trials testing novel drug treatments can fail due to inadequate selection of the right patient population. Drugs targeting other hormonal pathways have also been tested as potential novel treatments for schizophrenia. For example, the adrenal steroid hormone dehydroepiandrosterone was found to improve negative symptoms in medicated schizophrenia patients. Likewise, the oestrogen-related drug raloxifene led to reduced negative symptoms in postmenopausal female schizophrenia patients.

As inflammation was another major pathway affected in schizophrenia patients, another approach would be to use anti-inflammatory drugs to target those patients with perturbed immune or inflammatory biomarker profiles. To test this, recent studies investigated the effects of cyclooxygenase-2 (COX-2) inhibitors for treatment of schizophrenia symptoms. These studies found that patients treated with the antipsychotic amisulpride plus the COX-2 inhibitor celecoxib experienced greater improvements in negative symptoms compared to those patients treated with amisulpride alone. If these effects can be repeated in validation studies, a personalized medicine approach could also be used here by screening of patients for biomarkers related to immune status, followed by anti-inflammatory treatment of only those patients with high levels of inflammation-related proteins along with traditional antipsychotic treatment. One of the most well-known COX-2 inhibitors, aspirin, has also been tested successfully in combination with regular antipsychotic treatment for reducing the symptoms associated with schizophrenia spectrum disorders. Likewise, testing of non-steroidal anti-inflammatory drugs like ibuprofen, diclofenac and naproxen as an adjuvant with antipsychotics has already shown some promise for reducing schizophrenia symptoms along with the co-morbid inflammatory effects.

Can We Improve Treatment Response of Patients Using Biomarkers?

As mentioned earlier, one of the biggest challenges in improving the lives of schizophrenia patients is for physicians to find an effective drug treatment strategy. There are currently no firm guidelines and treatments are usually selected based on subjective criteria. This is not surprising as most current antipsychotic medications used in the treatment of schizophrenia were originally developed for treatment of other diseases. The finding that some patients also improvement in their psychiatric symptoms led to the increased use of these compounds in the treatment of psychiatric disorders. This is interesting as there is still not sufficient understanding of the underlying pathophysiology to know why these drugs were working. However, the problem is that treatment with these drugs and their later derivatives does not always lead to a good outcome for the patients in terms of a reduction in symptoms. Double blind randomized clinical trials (this is a trial in which neither the examiners nor the participants know who has received the active drug or a placebo) have found that response rates to typical first generation antipsychotic drugs range between 46 and 76 % of the patients showing a good response. Of course this also depends on the trial length and the response criteria used by the clinicians. Similar responses have also been found in clinical trials involving patients who have received the newer (atypical) types of antipsychotic drugs. For example, a 6-week clinical trial which compared the first generation drug haloperidol with the second generation drug risperidone found response rates of 56 % and 63 %, respectively, as shown by a significant reduction of PANSS scores. However, other trials found lower response rates for both of these drugs but a higher response rate of 67 % for the second generation compound olanzapine. The stage of the disease appears to play an important role in how well patients respond. The response rates in treatment-resistant patients are generally lower than those for first-episode schizophrenia patients as these can range from 30 to 60 %. These findings suggest that a suboptimal response to antipsychotics is a frequent occurrence. This explains why in current practice there is frequent switching and changing of medications and an overall therapeutic delay in tempering or eliminating psychotic symptoms. This can have a range of deleterious consequences in the lives of the patients, as well as knock-on effects on their families and friends, and on the work force, society and the healthcare systems in general. These factors alone highlight the critical need to identify better ways of treating these patients.

Although the application of blood based tests is becoming more and more common in clinical fields such as oncology, the use of biomarker tests for improved diagnosis and treatment response have still not emerged in the area of psychiatric disorders. However, a few pilot studies have been carried out over recent years which indicate that there is considerable promise in this endeavour. A study in 2006 showed that high baseline levels of serum cortisol and elevated cortisol/dehydroepiandrosterone ratios could be used to predict response to antipsychotic therapy at 2 and 4 weeks of treatment. Two studies were performed which measured the levels of serum biomarkers before commencement of treatment in 77 schizophrenia patients. The platform used for this analysis was the multiplex immunoassay technique described in the previous chapter. The main aim of this study was to determine whether any markers could be used to predict general response and even side effects to antipsychotic treatments given for 6 weeks to first onset schizophrenia patients. The analysis showed that baseline levels of insulin correlated with improvement in PANNS negative symptom scores at the end of the 6 week treatment period. Interestingly, lower insulin levels correlated with the best improvements while patients with higher insulin levels at the start of treatment showed little or no improvement. Also, weight gain and insulin resistance are well known side effects of antipsychotic treatment and a subsequent analysis of the same samples found that the baseline levels of 9 serum molecules were significantly associated with the increase in body weight. This consisted of two lipid transport molecules, three cytokines, two pituitary hormones, and two growth factors. In addition, the insulin/glucose ratio was increased by approximately 40 % between the baseline and 6 week follow up period, suggesting that the weight gain may have been associated with increased insulin resistance. The gaining of body weight associated with antipsychotic therapy is normally experienced within the first few months and may continue with ongoing treatment. Some factors such as younger age and lower initial body mass indices have been linked to treatment-induced weight gain. It is tempting to speculate that such patients show the greatest improvement as they have a higher capacity to become insulin resistant and gain weight.

Taken together, these findings suggest that insulin signalling and weight gain may play a role in the response to antipsychotic treatment. The association between insulin resistance, weight gain and therapeutic response has led many researchers to believe that these outcomes are an unavoidable side effect and may even be essential for clinical benefit. However, the results of the recent studies mentioned above (involving co-administration of antipsychotics and anti-diabetic drugs) have suggested that weight gain may not be a specific physiological factor that can explain the mechanism underlying this relationship. In fact, this study showed that the joint treatment approach actually led to improvement of symptom scores without the accompanying weight gain.

Development of Biomarker Tests to Detect Schizophrenia Before Disease Onset

Two recent studies have shown that it may be possible to identify individuals at risk of developing schizophrenia through multiplex immunoassay profiling even years before the onset of the disease. The first of these was a project carried out by Perkins and colleagues as part of The North American Prodrome Longitudinal Study project. They carried out multiplex immunoassay analyses of blood plasma analytes and selected a combination 15 of these which best distinguished 32 persons with clinical high-risk symptoms who developed psychosis from 35 similar subjects who did not develop psychosis during a 2-year follow-up period. This resulted in a test with 91 % accuracy. In the second study, Chan and colleagues carried out a three step procedure. The first step involved analyses of five different cohorts of 127 first-onset schizophrenia patients and 204 controls which resulted in identification of a panel of 26 biomarkers that best discriminated patients from controls. Next, they validated this biomarker panel using two further cohorts of 93 patients and 88 controls, and this gave an accuracy of 97 % for schizophrenia detection. Finally, they tested the predictive performance of this panel for identifying at-risk individuals up to 2 years before the onset of psychosis using two cohorts of 445 at-risk individuals.

By also considering symptom scores (in this case Comprehensive Assessment of At-Risk Mental State [CAARMS] positive symptom scores) they were able to identify the population who would eventually develop schizophrenia with 90 % accuracy. Further development and validation both of these tests will aid clinicians in the identification of vulnerable patients early in the disease process, allowing more effective therapeutic intervention before overt disease onset.

Future Prospects in Schizophrenia Research

This chapter describes recent advances using proteomic biomarkers for improved diagnosis and classification of individuals with schizophrenia with the ultimate goal of providing them with better treatments for improved outcomes. The use of multiplex proteomic profiling analysis has presented us with a way of de-convoluting the wide spectrum of molecular pathways affected in the disease. This has led to a more integrative view of the biological pathways thought to be perturbed in schizophrenia, as a disease which affects multiple organ systems of the body and not just the brain. Many patients show perturbations in circulating molecules suggestive of metabolic abnormalities such as altered HPA axis function, high insulin levels or insulin resistance, and some show a dysfunctional immune system with elevated levels of inflammatory cytokines. The possibility that different patients have distinct alterations in either one but not both of these pathways suggests that schizophrenia may comprise different subtypes, at least at the molecular level. Therefore, improved classification of patients based on molecular profiling would enable stratification of these individuals prior to treatment. This would require adopting a phenotype-based approach by using specific readouts such as hormone imbalances and immune factors as a guide. This could result in more effective treatments with fewer side effects and, therefore, a lower rate of medication discontinuation.

Stratification of schizophrenia patients using proteomic biomarker profiles to deliver the correct treatments to the right patients is a form of personalized medicine. This has probably been shown best in the field of cancer. For example, the measured over-expression of the human epidermal growth factor receptor 2 helps to identify those women with breast cancer who are more likely to benefit from treatment with the monoclonal antibody Herceptin, which blocks the activity of this receptor (as described earlier). Similar efforts in the field of schizophrenia will hopefully lead to identification of novel therapeutic targets and to individualization of treatment strategies. This will increase the chances of positive therapeutic outcomes. In addition, new adjunctive drug treatment strategies could be developed which target co-morbidities such as metabolic syndrome or pro-inflammatory conditions, as evinced by the presence of specific biomarkers. Thus, considering schizophrenia as a whole body disease may eventually lead to better treatment approaches to improve the lives of patients with this distressing psychological illness.

Chapter 5
Progress for Better Treatment of Depression

Depression is a serious illness affecting a staggering one out of every five people in their lifetime. The disease is characterized by extreme low energy and mood, often combined with thoughts of worthlessness and suicidal ideations. There are many risk factors which we are now only just beginning to come to terms with. As far as treatment goes, there is still considerable room for improvement. As with most of the psychiatric disorders, the treatment of depression is often a subjective trial and error process involving administration and switching of drugs and drug doses multiple times until an adequate response is achieved. In many cases, this is never attained by using traditional antidepressant drugs alone and sometimes more drastic measures are required, such as administration of electroconvulsive therapy. This chapter describes how more successful treatment strategies can be achieved by simply improving our understanding of the underlying disease mechanisms. Several studies in the area of depression have now shown that particular pathways in the body are perturbed, including neurotransmitters, the immune system and hormonal networks. Development of biomarkers and disease models reflecting these pathways could lead to our increased understanding of the disease and, consequently, the design of better drugs to treat patients with this debilitating disorder.

Why Is It Difficult to Diagnose Depression?

Major depressive disorder is the most common psychiatric illness and is a leading cause of disability and other health problems throughout the word. Depression can be manifested in several distinct ways and it can have substantial financial costs to society. The main symptoms are notably a depressed mood, changes in sleep duration or sleep patterns, a feeling of low energy levels, loss of interest in multiple aspects of life, changes in appetite, gaining or losing weight, feelings of worthlessness, general despair and an inability to concentrate or focus on details. In general medical care, clinical depression has been both under and over diagnosed. For

© Springer International Publishing AG 2017
P.C. Guest, *Biomarkers and Mental Illness*,
DOI 10.1007/978-3-319-46088-8_5

example, less than 50 % of depressed individuals were identified correctly by their physicians in a study that investigated the 5 year mental disorder recognition rates in general practice. This was confirmed by a large compilation of several studies (known as a meta-analysis) which analyzed data from more than 50,000 patients. This confirmed the above mentioned findings by showing that general practitioners could correctly identify depression in only 47 % of the cases. The problem of not correctly identifying depression in the primary care setting can have detrimental effects on the course and outcome of the illness as in these cases the patients are not likely to receive treatment, at least not one that they would benefit from. Interestingly, the correct diagnosis of depression in individuals varies widely in different countries, ranging from as low as 19 % in Japan to as high as 74 % in Chile. It is not clear why this occurs but it could be due to differences in medical practice in these countries or it could be down to cultural reasons.

There may be several reasons why depression is under-diagnosed. One study found that around two-thirds of depressed patients actually go to their general practitioners with complaints which are obviously physical, such as low energy, sleep disturbances and various pains. In these scenarios, doctors might overlook that these symptoms could be linked or even caused by a mental health issue. Also, the patients themselves might not like to be diagnosed with depression and resist this if they believe that their symptoms are the result of physical causes. The problem of a physician not reaching a correct diagnosis, or a patient not accepting it, could also be serious in some cultures in which a mental illness is associated with negative stigma as it can lead to social exclusion. Conversely, over-diagnosis can also be a problem as this could lead to negative effects on the patient such as an unnecessary drug treatment which comes with a strong risk of adverse side effects. Interestingly, a cross-sectional survey of 8796 members of the general populations in Belgium, France, Germany, Italy, the Netherlands and Spain found that 13 % of individuals diagnosed with depression by general practitioners did not actually have any mental disorder.

Finally, limitations in time and restrictions of resources are an ever increasing problem in primary care. Therefore, these factors are likely to be a main cause of under-diagnosis of depression. The average time that a patient spends with the physician in primary care is around 6 min, which is obviously not stringent enough when dealing with complex conditions such as psychiatric disorders.

Diagnostic Tools

DSM-5

There are three criteria for major depressive episode, termed A through C. For criterion A, five or more of the following symptoms must be present during the same 2 week period nearly every day and represent a change from previous functioning:

1. Depressed mood most of the day
2. Markedly diminished interest or pleasure in activities most of the day

3. Significant weight loss or gain or decrease or increase in appetite
4. Insomnia or hypersomnia
5. Psychomotor agitation or retardation
6. Fatigue or loss of energy
7. Feelings of worthlessness or inappropriate guilt (may be delusional)
8. Diminished ability to think or concentrate, or indecisiveness
9. Recurrent thoughts of death or suicidal ideation

In addition, at least one of the symptoms should be either (1) depressed mood or (2) loss of interest or pleasure. For criterion B, the symptoms should cause clinically significant distress or impairment in social, occupational or other important areas of functioning. For criterion C, the symptoms should not be due to the direct physio-logical effects of a substance, such as a drug of abuse or medication, or a general medical condition, including hypothyroidism.

Hamilton Depression Rating Scale (HAM-D)

The HAM-D test has been used for several years to calculate the severity of depression before, during and after treatment. Although the test lists 21 items, scoring is based only on the first 17, with higher scores indicating a more severe case of depression.

1. Depressed Mood (sadness, hopeless, helpless, worthless)

   0=absent
   1=these feeling states indicated only on questioning
   2=these feeling states spontaneously reported verbally
   3=communicates feeling states non-verbally—i.e. through facial expression, posture, voice, and tendency to weep
   4=patient reports virtually only these feeling states in his spontaneous verbal and non-verbal communication

2. Feelings of Guilt

   0=absent
   1=self reproach, feels he has let people down
   2=ideas of guilt or rumination over past errors or sinful deeds
   3=present illness is a punishment. Delusions of guilt
   4=hears accusatory or denunciatory voices and/or experiences threatening visual hallucinations

3. Suicide

   0=absent
   1=feels life is not worth living
   2=wishes he were dead or any thoughts of possible death to self
   3=suicidal ideas or gesture
   4=attempts at suicide (any serious attempt rates 4)

4. Insomnia (Early)

   0=no difficulty falling asleep
   1=complains of occasional difficulty falling asleep—i.e. more than 1/2 h
   2=complains of nightly difficulty falling asleep

5. Insomnia (Middle)

   0=no difficulty
   1=patient complains of being restless and disturbed during the night
   2=waking during the night—any getting out of bed rates 2 (except for purposes
      of voiding)

6. Insomnia (Late)

   0=no difficulty
   1=waking in early hours of the morning but goes back to sleep
   2=unable to fall asleep again if he gets out of bed

7. Work and Activities

   0=no difficulty
   1=thoughts and feelings of incapacity, fatigue or weakness related to activities;
      work or hobbies
   2=loss of interest in activity, hobbies or work—either directly reported by
      patient, or indirect in listlessness, indecision and vacillation (feels he has to
      push self to work or activities)
   3=decrease in actual time spent in activities or decrease in productivity
   4=stopped working because of present illness

8. Retardation (Psychomotor) (slowness of thought and speech; impaired ability
   to concentrate; decreased motor activity)

   0=normal speech and thought
   1=slight retardation at interview
   2=obvious retardation at interview
   3=interview difficult
   4=complete stupor

9. Agitation

   0=none
   1=fidgetiness
   2=playing with hands, hair etc.
   3=moving about, cannot sit still
   4=hand wringing, nail biting, hair-pulling, biting of lips

10. Anxiety (Psychological)

    0=no difficulty
    1=subjective tension and irritability
    2=worrying about minor matters
    3=apprehensive attitude apparent in face or speech
    4=fears expressed without questioning

11. Anxiety (Somatic): Physiological concomitants of anxiety (i.e. effects of autonomic overactivity, "butterflies", indigestion, stomach cramps, belching, diarrhea, palpitations, hyperventilation, paresthesia, sweating, flushing, tremor, headache, urinary frequency).

0=absent
1=mild
2=moderate
3=severe
4=incapacitating

12. Somatic Symptoms (Gastrointestinal)

0=none
1=loss of appetite but eating without encouragement from others. Food intake about normal
2=difficulty eating without urging from others. Marked reduction of appetite and food intake

13. Somatic Symptoms (General)

0=none
1=heaviness in limbs, back or head, backaches, headaches, muscle aches, loss of energy and fatigability
2=any clear-cut symptom rates 2

14. Genital Symptoms (symptoms such as: loss of libido, impaired sexual performance, menstrual disturbances)

0=absent
1=mild
2=severe

15. Hypochondriasis

0=not present
1=self-absorption (bodily)
2=preoccupation with health
3=frequent complaints, requests for help etc.
4=hypochondriacal delusions

16. Loss of Weight (Rating by History)

0=no weight loss
1=probably weight loss associated with present illness
2=definite (according to patient) weight loss
3=not assessed

17. Insight

0=acknowledges being depressed and ill
1=acknowledges illness but attributes cause to bad food, climate, overwork, virus, need for rest etc.
2=denies being ill at all

18. Diurnal Variation
A. Note whether symptoms are worse in morning or evening (if no diurnal variation, mark 0)

   0 = no variation
   1 = worse in A.M.
   2 = worse in P.M.

B. When present, mark the severity of the variation (mark 0 if no variation)

   0 = none
   1 = mild
   2 = severe

19. Depersonalization and Derealization (feelings of unreality, nihilistic ideas)

   0 = absent
   1 = mild
   2 = moderate
   3 = severe
   4 = Incapacitating

20. Paranoid Symptoms

   0 = none
   1 = suspicious
   2 = ideas of reference
   3 = delusions of reference and persecution

21. Obsessional and Compulsive Symptoms

   0 = absent
   1 = mild
   2 = severe

   Scoring (maximum = 50)
   0–7 = normal
   8–13 = mild depression
   14–18 = moderate depression
   19–22 = severe depression
   ≥23 = very severe depression

Why Is It So Difficult to Treat Depression?

One reason why major depressive disorder can be so devastating to the lives of patients is because the associated treatment periods are likely to be long and arduous. The current treatment guidelines recommend the use of a waiting period, which allows the clinician to determine if a particular antidepressant medication will be effective. In this case effective means that the treatment will result in an adequate response (greater than a 50 % reduction in depressive symptoms). On average, it can take 4 weeks before some response to antidepressant treatment is observed and approximately 12 weeks to achieve remission. To make things worse, the patients who show no response can then be subjected to one or more additional periods of treatment, during which increasing dosages of the same drug are administered or different classes of antidepressant drugs are tested. This trial and error process can take 1 year or even longer before a satisfactory state of recovery is achieved. Approximately one quarter of the patients who do not improve with the first antidepressant treatment stop taking the medication, and this often happens within the first two weeks. As treatment time goes on, more and more patients stop taking their medication, with almost half coming off the drug during the first month and around three quarters stopping their therapy by 3 months.

To date, more than 20 antidepressant drugs for major depressive disorder have been approved by the FDA. However, the associated response and remission rates for all of these drugs have so far been poor. A meta-analysis of randomized, double-blind clinical trials found that the response rate for antidepressants was only around 50 % and, surprisingly, it was almost 40 % for the placebo effect. One study of more than 4000 patients found that only one third achieved remission after 12 weeks of treatment with the initial antidepressant and the remainder required multiple treatment stages. The consequences of a lengthy treatment regime can be serious for patients since this will naturally result in extended periods of time that they spend in the depressed state. This leads to increased disability and health care costs along with impaired productivity in society or the work place. Taken together, these factors spell out the need to identify biomarkers which could be used to predict treatment response, and thereby reduce the duration of ineffective treatment and improve patient compliance.

What Is Going on in the Brains of Depressed Patients and Why Does It Take Antidepressants So Long to Work?

Antidepressants such as SSRIs are usually the first line of therapy, although the mechanisms by which they exert their therapeutic effects are poorly understood. For example, it is not known why several weeks of antidepressant treatment are required before any clinical improvements are seen. Recent theories to explain this lag include the hypothesis that antidepressants act to restore neurogenesis (this is the process of making new neurons) and presumably alter synaptic connectivity in the hippocampus. However, because the majority of clinically used antidepressants target neurotransmitter systems, it is difficult to determine whether changes in brain function after chronic administration of these drugs reflect their psychotherapeutic properties, or simply result from alteration of specific neurotransmitter receptor signalling pathways.

One study was carried out in 2004 which confirmed that the lag could be due to the time required for enabling rewiring of the brain—known as synaptic remodelling. This study used the proteomic technique called 2DE using fluorescent dyes to

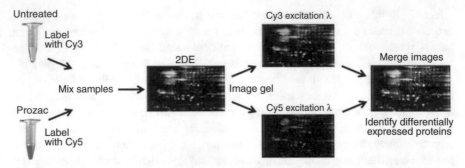

Fig. 5.1 Diagram showing 2DE with fluorescent CyDyes to identify proteins expressed at different levels. Proteins in the extracts under comparison are labelled with CyDyes that fluoresce at distinct wavelengths. In this case, these were brain proteins from animals treated with either a saline control (labelled with Cy3) or Prozac (Cy5). The labelled samples are then mixed together and subjected to 2DE. After running the gel in both dimensions (see Chap. 3), the gel is imaged at the wavelengths specific for each CyDye, using different lasers for excitation. This produces images of the two different proteomes which consist of protein spots that are either *green* or *red* in colour. After this, the images are overlaid such that unchanging proteins appear *yellow* and proteins that are present at a higher level in the Cy3 labelled sample appear green and those that are higher in the Cy5 sample are coloured *red*

label brain proteins from fluoxetine (Prozac)-treated and untreated guinea pigs and the investigators looked for the proteins that changed after 2 weeks (Fig. 5.1). This showed that the antidepressant treatment led to alterations in the expression of several key components of synaptic structure and function in the cerebral cortex.

The main effect showed a decrease in the levels of neurofilament (NF) proteins in the animals treated with Prozac. The NF family of proteins has four members termed light (NF-L), medium (NF-M), heavy (NF-H) and alpha-internexin (αINX). These are localized to neurons in the dendrites, axons and presynaptic terminals (see Chap. 1) and numerous studies suggest that the NFs undergo down-regulation and disassembly during the dynamic cytoskeletal remodelling involved in new synapse formation. In fact, low levels of NFs are expressed in developing neurons during the dynamic growth phase of axonal and dendritic architecture with a switch to high NF levels during stabilization of the formed processes.

To follow up on this idea, the authors carried out electron microscopy analysis which provided visual evidence for an increase in the number of synapses undergoing remodelling (Fig. 5.2). This revealed a twofold increase in the number of split synapses—a morphological change thought to be indicative of synapses in the process of making new connections. It is thought that split synapses represent dynamic structures, which undergo increased recycling of synaptic membranes and associated proteins as the neurons react to increasing or decreasing synaptic strength. Taken together, these findings suggest that the characteristic delay in the onset of efficacy of antidepressant therapies reflects the time required to increase the rate of synaptic remodelling in the appropriate neural networks.

Untreated   Prozac

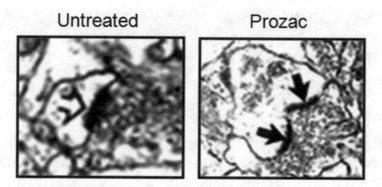

Fig. 5.2 Electron microscopy analysis showing a twofold increase in the number of guinea pig neurons with a split synapse configuration after Prozac treatment (*solid arrows*) compared to those with a single synapse structure in untreated animals (*open arrow*)

This hypothesis is supported by previous studies in which chronic electroconvulsive seizure treatment also decreased expression of NF proteins in the hippocampus and by a study demonstrating reversible changes in dendritic spine density in an animal model of depression. This is interesting because recent studies suggest that depression may be associated with disruption of the mechanisms that regulate neuronal survival, networking and plasticity in the brain and antidepressants could reverse these effects by increasing neurogenesis and neuronal plasticity. Antidepressants have also been shown to increase neurotrophin-mediated neuronal growth and excitatory neurotransmitter-mediated synaptic plasticity in the hippocampus and other brain structures. Other studies have shown that chronic antidepressant treatment prevented atrophy of dendrites in the hippocampus and partially reversed the long-term changes of hippocampal synaptic plasticity in stressed rats. Taken together, these effects indicate that antidepressants may work by stimulating neuronal growth, new synaptic connections and survival.

Can We Detect Depression Using Biomarkers in the Blood?

Again, the key point to remember here is that the brain does not work alone. Most of its functions are regulated by communication with hormones, growth factors and inflammation-related molecules, which circulate throughout the body in the bloodstream. As with schizophrenia in the previous chapter, changing patterns of molecules in the blood offers a means of monitoring brain function in patients with major depressive disorder (Fig. 5.3). There is even the possibility that this could be used to provide a way of distinguishing individuals who suffer from different subtypes of this disorder.

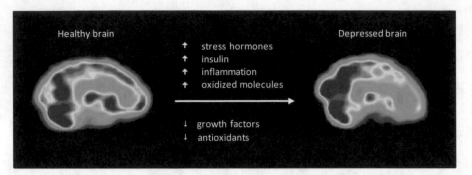

Fig. 5.3 Classes of biomarkers affected in depression. *Up arrows* indicate that the biomarker is increased and *down arrows* indicate the opposite. The figure also shows a SPECT imaging analysis depicting decreased blood flow in frontal brain regions of a depressed brain compared to a healthy brain

Stress Hormones

Investigations have shown that circulating ACTH and cortisol levels are increased in major depressive disorder patients with the melancholia subtype compared with non-melancholic depressed patients and healthy controls. This indicates that the HPA axis is activated in depression, an idea which has gained support from a number of different studies. Also, some patients in the early phases of depression show an enhanced responsiveness to stress, again supporting the likelihood of changes in regulation of the HPA axis.

Other Hormones

In line with the effects on stress hormones, researchers have noted effects on other hormone systems of the body. Many patients with depression develop glucose intolerance and hyperinsulinemia, suggestive of insulin resistance. One investigation found that insulin signalling was significantly impaired in depressed patients compared to controls and they found that treatment with antidepressants resulted in a significant improvement in insulin sensitivity. Again, there seems to be a link between psychiatric disorders and metabolic conditions such as type II diabetes. The link between insulin resistance and impaired brain function can be explained since inadequate function of this pathway can lead to poor metabolism in neuronal cells and consequential impairment of central nervous system responses, such as mood, motivation, emotions and cognition. Effects on other metabolism-related hormones have also been found to occur in depressed patients. One study showed a significant correlation between the levels of the lipid metabolism-related hormone leptin with depressed mood and sleep disturbances in female patients, although this only occurred in those with a normal body weight. No such correlations were found in overweight female patients or in males of any body mass index. Leptin is known for its role in lipid metabolism and changes in this hormone in depression could therefore be linked to the insulin resistance issue described above. Two other studies found that the levels of adiponectin, another hormone involved in regulation of lipid metabolism, are also reduced in patients with major depressive disorder, although these investigations did not make a distinction between males and females or other factors such as body mass index. The effects on hormones involved in lipid metabolism are in line with other studies which have shown perturbations in the levels of circulating lipids like cholesterol in depressed patients. Another study found that thyroid stimulating hormone (TSH) and thyroxine (T4) levels may also be correlated with the severity of depressive symptoms, which may not be surprising. Given the role of the thyroid hormones in setting the body's thermostat for regulation of such vital functions as the basal metabolic rate, protein synthesis, long bone growth, maturation of neurons, and sensitivity to catecholamines, changes in this pathway might explain the effects on other hormones and growth factors found in depression and other psychiatric disorders.

Growth Factors

Several studies have found altered levels of BDNF in the blood of depressed patients, although these findings have not always been consistent. As the name implies, this protein is involved in the growth, proper function and survival of neuronal cells in the brain. One study found lower BDNF levels in un-medicated depressed patients and showed that these were negatively correlated with HAM-D scores. Thus, the findings of low BDNF levels in depression, suggests that high levels of this growth factor are required for normal brain function. Consistent with this possibility, studies of depressed patients have also identified changes in the levels of vascular endothelial growth factor (VEGF), fibroblast growth factor 2 (FGF-2) and angiotensin-converting enzyme (ACE), which all have effects on blood circulation in the brain through regulation of both peripheral and central vascularisation. This has also been supported by another study which demonstrated increased levels of ACE activity in serum from major depressive disorder patients concomitant with readings of higher blood pressure in the same individuals.

Inflammation

A large number of studies have also been carried out which have implicated altered inflammatory processes in the pathogenesis of depression. A multiplex immunoassay study reported changes in the levels of monocyte-, macrophage-and cytokine-related proteins in the blood of patients with major depression compared to healthy control subjects. Other studies which used specific immunoassays to target single proteins confirmed increased levels of both proinflammatory and anti-inflammatory cytokines in depressed patients. Therefore, this latter study showed a mixed change in molecules that either promoted or reduced inflammation. Similarly, increased levels of other inflammatory biomarkers have been identified in depression including several acute phase response proteins. These are circulating proteins which are altered in response to inflammation that follows an injury, infection or disease. Along these same lines, a recent investigation found that elevated serum levels of the cytokine interleukin 5 are associated with an increased risk of major depressive disorder, and another study found that a pattern of high eotaxin and low monocyte chemotactic protein 1 and RANTES (also known as the lengthy titled protein, "regulated-on-activation normal T cell expressed and secreted") occurred in depressed females who had contemplated suicide. Further work should obviously be carried out in this area taking into account the high percentage of depressed patients who consider or actually complete suicide. Still, it is not known whether such changes in inflammation-related molecules are causative or a consequence of the disease.

Other evidence consistent with altered inflammation in major depression comes from studies which investigated the effects of exposure to inflammatory challenges such as application of lipopolysaccharide and cytokines (these could basically be considered as irritants). An injection with lipopolysaccharide results in inflammation in both the periphery and the central nervous system, as shown by the long-lasting increase in the levels of the pro-inflammatory protein tumour necrosis factor alpha (TNFα). Furthermore, several investigations have shown that treatments which induce increased production of pro-inflammatory cytokines, can also lead to behavioural changes in people which are similar to those found in depression. In this way, treatment of cancer or hepatitis C with interferon-α can cause depression in a high proportion of the patients. All members of the interferon family are antiviral agents which modulate the immune and anti-inflammatory systems by initiating a response to an infection (normally viral).

Oxidative Stress

Another pathway which may be altered in depression is the regulation of oxidative stress. This is an imbalance between the body's ability to detoxify oxidized molecules (know as reactive oxygen species) and the manifestation of the reactive molecules themselves. The entire pathway is known as the redox state (for reduction/oxidation) as the system attempts to prevent or repair the oxidative damage through the process of reduction. Disturbances in the redox state in all tissues can result in toxicities via production of peroxides and free radicals that cause damage to proteins, lipids, and DNA. This may be an over simplification but the process could be compared to a rusted automobile. Interestingly, researchers have used redox biomarkers to distinguish patients with different sub-types of depression. They found that it may be possible to sub-classify major depression patients into either early onset or chronic subclasses using the circulating levels of two molecules which normally prevent oxidative damage (superoxide dismutase and glutathione peroxidise) and one which indicates that such damage has already occurred (malondialdehyde). The study showed a significant decrease in serum superoxide dismutase and glutathione peroxidase along with an increase in malondialdehyde in recurrent depressive patients in comparison to first episode patients. This indicated a shift towards an increase in oxidative stress in depressive patients. Again, it is not clear whether such effects are a cause or result of the disorder.

Can We Develop a Blood Test for Major Depressive Disorder?

Researchers recently described the development of a multiplex biomarker test comprised of nine serum molecules which could be used to distinguish patients with major depressive disorder from healthy control subjects. The test had a sensitivity of 92 % for identification of true patients and a specificity of 81 % for identification of true controls. However, as with all biomarker tests, these results are merely promising and considerable further work is needed before such a test can be implemented in a clinical or laboratory setting. It will be important to follow up this research with validation testing using new cohorts and at multiple clinical sites. It is also essential that further research is also aimed at determining the specificity of this test for depression through studies carried out using patients with other psychiatric disorders. This is important due to the notorious overlap of psychiatric diseases described earlier which may make them difficult if not impossible to separate at the level of blood-based biomarkers. These further steps will help to avoid the scientific community jumping to premature conclusions, which has already led to considerable confusion in this field. Nevertheless, the above results suggest the presence of robust circulating biomarkers in patients with depression and further studies in this area could lead to the development of new blood tests which could be used by primary care physicians, since time and resources for diagnosis of depression are normally limited in this setting.

Are There Any Novel Treatments on the Horizon for Depression?

Considering the fact that major pharmaceutical companies have virtually abandoned the search for novel treatments, mainly due to lack of understanding of the aetiology and progression of the disease, knowledge of the molecular pathways involved in depression could help get the ball rolling again with regards to research into new treatment options. One of the most promising areas for treatment of depression revolves around the potential positive effects of exercise. A study from 2005 revealed that antidepressant treatments and physical exercise can relieve symptoms of depression, up-regulate the release of neuronal growth factors and increase neurogenesis in the hippocampal region of the brain. The researchers measured the effects on cell proliferation by measuring the levels of neurotrophic factors such as BDNF and dynorphin in rats that either had or did not have access to running wheels.

Amazingly, wheel running appeared to have an antidepressant effect on the rats, as shown the results of a standard rat behavioural measurement called the Porsalt swim test (PST). This test is based on the assumption that an animal will try to escape a stressful experience and, when escape is impossible, it eventually stops trying and gives up. The test begins by placing the animal in an inescapable glass cylinder containing water. Most animals attempt escape by swimming but will eventually "give up" and float on the surface of the water. An animal that gives up quickly is thought to display characteristics similar to human depression. The researchers in the above study showed that the animals that attempted escape for longer time periods before giving up also had increased production of new neurons along with higher BDNF and dynorphin levels in the hippocampus. From this, the authors supported the ideas of other researchers who suggested that depression most likely involves disruption of neurogenesis in the hippocampus, and they proposed the new idea that physical activity might have antidepressant effects by reversing this. More recently, Huang and co-workers carried out a study in rodents which resulted in identification of common neuronal and gene expression changes in response to both antidepressant drugs and exercise. This suggested that there was a shared mechanism underlying these effects.

Intensive research into the molecules responsible for the association between exercise and depression has been carried out for over a decade. One of the most significant molecules implicated in this association is BDNF. A study by Suiciak and co-workers demonstrated that BDNF had antidepressant-like properties, by directly injecting this molecule into the midbrains of two animal models of depression (Fig. 5.4). Also, genetic knockdown (a technique whereby transcription and translation of a gene sequence into the encoded protein is experimentally dimin-

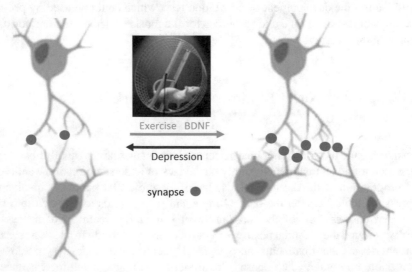

Fig. 5.4 Exercise leads to increased levels of nerve growth factors such as BDNF, which in turn increases synaptic connectivity and thereby alleviates some symptoms of depression

ished) of BDNF resulted in depression-like behaviour. A recent study confirmed the association between BDNF and depression through studies which exposed the animals to an acute stressor. Interestingly, the animals responded in one of two ways to the stress by showing either vulnerability or resistance to development of depressive characteristics. Most interestingly, the BDNF levels were lower in those rats that exhibited depressive behaviour after application of the stressor compared to their counter parts that showed non-depressive behaviour. These studies demonstrate the potential use of the BDNF pathway as a predictive biomarker and a novel drug target for treatment of depression. Furthermore, and perhaps most importantly, this highlights the need for further research into the potential positive effects of exercise on individuals with mental disorders such as depression.

Future Prospects in Depression Research

This chapter presented state-of-the-art research using circulating blood-based biomarkers that may someday help to improve the diagnosis and classification of individuals with depression. As with schizophrenia, this has shown that there is a vast array of molecular pathways in the periphery and the brain which are affected in the disease. Many depressed patients have changes in circulating molecules which indicate changes in metabolism, insulin signalling, the HPA axis, growth factors, redox signalling and the immune and inflammatory systems. However, note the similarity of these changes to those seen in schizophrenia. This again raises the question of whether it will be possible to distinguish these disorders based on blood-based biomarker tests. Of course this is critical to ensure that patients are administered the right treatments for increasing the chances of a positive outcome. Finally, the evidence suggests that exercise helps to alleviate depressive symptoms via the regulation of two major pathways in the brain: synaptic plasticity and neurogenesis. Could this mean that exercise will be a cure all for depression? The answer is: probably not. Severe cases of depression are likely to more resistant. Furthermore, by definition, many depressed patients show lack of energy or movement and thus it will most likely not be possible to motivate individuals who suffer with such symptoms. Nevertheless, significant further research is warranted in this area, considering the potential benefits for people suffering from this devastating psychiatric illness, which appears to affect so many people worldwide.

Chapter 6
The Special Case of Bipolar Disorder

Bipolar disorder (BD) is one of the most difficult psychiatric disorders to investigate given the wide variation in symptoms, the similarity of these symptoms to those seen in other psychiatric disorders and the problems induced by cycling of mood states between the extremes of mania and depression. These cycles can occur over short time periods or they can even take several years to come about. One of the most confusing factors is that patients with bipolar disease often present first with symptoms that are indistinguishable from depression. Taken together, these factors make identification of individuals with this disorder difficult and, as such, diagnosis can often be delayed or even wrong. This also makes it difficult to treat patients since incorrect or delayed therapy could make these patients worse or even precipitate violent mood swings, or cause a shift in mood state or the development of other medical complications due to medication side effects. There have now been a number of biomarker-based investigations which have been carried out in attempts to increase our knowledge of this disease to help tackle the complications of diagnosis and shed new light on potential newer and better treatments. This chapter reviews the behavioural characteristics of bipolar disorder, followed by outlining what the biomarker studies have told us about what is going on at the level of the pathophysiology. Again, this appears to involve peripheral systems such as alterations in insulin signalling and perturbed inflammatory pathways. Finally, the potential areas for novel and improved treatment approaches are outlined which have been guided by these investigations.

What Is BD Exactly?

The short answer is that we do not know. BD is a complex and devastating disease which causes suffering to patients and their families and can have knock on effects on many aspects of society. It is characterized by episodic changes in mood, which can cycle mostly between mania and depression (Fig. 6.1).

© Springer International Publishing AG 2017
P.C. Guest, *Biomarkers and Mental Illness*,
DOI 10.1007/978-3-319-46088-8_6

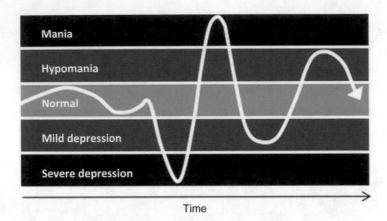

Fig. 6.1 Scheme showing the changing mood states of bipolar disorder over time

This cycling is often separated by periods of euthymic or so-called *normal* mood states. Mania is the defining feature of BD and is recognized as a distinct period of elevated or irritable mood that can take the form of euphoria and last for more than 1 week. A manic phase is often associated with an apparent increase in energy, a decreased need for sleep (less than 3 h per night in some cases), racing thoughts, poor attention span, increased risk taking, increased self importance and a heightened sex drive. Manic patients can also show signs of psychosis similar to those seen in schizophrenia. However, a mixed mood state can also occur which can be even more dangerous for the welfare of the patient. This is characterized by simultaneous presence of both manic and depressive symptoms, leading to increased unpredictability. Patients in this confused state frequently experience additional symptoms such as agitation, anxiety, guilt, impulsiveness, ideas of suicide and paranoia.

The effects of this disorder on society can also be devastating. The loss of years due to disability associated with BD has been estimated to be half of that due to diabetes, despite the fact that BD has a sixfold lower prevalence. To complicate matters, the pathological basis of BD is still only poorly understood. However, genetic and epidemiological studies have provided some insights. For example, relatives of BD patients are more likely to develop a mood disorder indicating that one or more genetic factors may be involved. In addition, genetic association studies have added new insights on increasing our understanding of the molecular nature of BD and imaging studies have found that the brains of BD patients show abnormal function in the prefrontal cortex, hippocampus and amygdala emotion-processing circuits. This also appears to involve changes in energy consumption in multiple brain regions, coinciding with the switches in mood states from depression to mania and to depression again (Fig. 6.2). Furthermore, there appears to be an over-activity of ventral striatal-ventrolateral and orbitofrontal cortex reward-processing pathways on the left side of the brain. These functional abnormalities appear to be associated with loss of grey matter in the prefrontal and temporal lobes, amygdala and

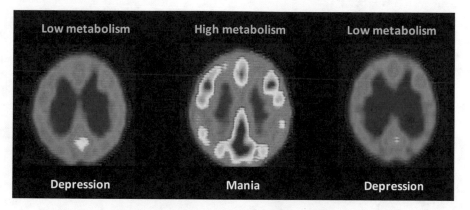

Fig. 6.2 PET imaging showing changes in brain energy consumption with the switch between mood states. The *blue* and *yellow/red* colours indicate areas of low and high energy consumption, respectively

hippocampus, along with decreased white matter tracts that normally connect the prefrontal and subcortical regions. Therefore, BD could result from a disruption of the normally co-ordinated brain circuitry as found with other psychiatric disorders like schizophrenia.

Diagnostic Tools

DSM-5

Bipolar disorders have been given their own separate description in DSM-5, although these were previously combined with depressive disorders in a chapter on mood disorders. The main criteria for diagnosis of bipolar mania or hypomania episodes should include the presence of markedly expansive, elated or irritable moods, although a new emphasis has been placed on changes in activity and energy levels. The diagnosis of "bipolar not otherwise specified" has been replaced by one described as "other specified bipolar and related disorder" to take into account people who may have a history of depression and meet all the criteria of hypomania, except the duration. In addition to the typical diagnoses of bipolar mania, hypomania, and depression, further information about the mood can be denoted with a "specifier" that further clarifies the course, severity or special features of the disorder or illness. The "with mixed features" specifier describes a mood that contains both mania/hypomania and depressive symptoms simultaneously. The specifier "with anxious distress" denotes patients suffering from anxiety symptoms that are not part of the bipolar diagnostic criteria.

Young Mania Rating Scale

The Young Mania Rating Scale (YMRS) is frequently used to assess symptoms of mania. The test has 11 items based on the subjective self report of the patients or their condition over the previous 48 h as well as on additional information based on clinical observations made during the interview. The YMRS follows the style of the HAM-D with each item given a severity rating. Seven of the items are given a 0–4 scale although the first 4 are given double this (0–8) to compensate for the possibility of poor cooperation from severely ill patients. Typical YMRS baseline scores can vary. The strengths of the YMRS are its brevity, its widely accepted use and its ease of administration. A score of ≥30 indicates severe illness and 1 of ≤12 indicates remission of symptoms.

1. Elevated Mood

   0=absent
   1=mildly or possibly increased on questioning
   2=definite subjective elevation; optimistic, self-confident, cheerful, appropriate to content
   3=elevated, inappropriate to content, humorous
   4=euphoric, inappropriate laughter, singing

2. Increased Motor Activity-Energy

   0=absent
   1=subjectively increased
   2=animated, gestures increased
   3=excessive energy, hyperactive at times, restless (can be calmed)
   4=motor excitement, continuous hyperactivity (cannot be calmed)

3. Sexual Interest

   0=normal; not increased
   1=mildly or possibly increased
   2=definite subjective increase on questioning
   3=spontaneous sexual content, elaborates on sexual matters, hypersexual by self-report
   4=overt sexual acts (toward patients, staff or interviewer)

4. Sleep

   0=reports no decrease in sleep
   1=sleeping less than normal amount by up to 1 h
   2=sleeping less than normal by more than 1 h
   3=reports decreased need for sleep
   4=denies need for sleep

5. Irritability

   0=absent
   2=subjectively increased

4 = irritable at times during interview, recent episodes of anger or annoyance on ward
6 = frequently irritable during interview, short, curt throughout
8 = hostile, uncooperative, interview impossible

6. Speech (Rate and Amount)

0 = no increase
2 = feels talkative
4 = increased rate or amount at times, verbose at times
6 = rushed, consistently increased rate and amount, difficult to interrupt
8 = pressured, uninterruptible, continuous speech

7. Language-Thought Disorder

0 = absent
1 = circumstantial, mild distractibility, quick thoughts
2 = distractible, loses goal of thought, changes topics frequently, racing thoughts
3 = flight of ideas, tangentiality, difficult to follow, rhyming, echolalia
4 = incoherent, communication impossible

8. Content

0 = normal
2 = questionable plans, new interests
4 = special project(s), hyper-religious
6 = grandiose or paranoid ideas, ideas of reference
8 = delusions, hallucinations

9. Disruptive-Aggressive Behaviour

0 = absent, cooperative
2 = sarcastic, loud at times, guarded
4 = demanding, threats on ward
6 = threatens interviewer, shouting, interview difficult
8 = assaultive, destructive, interview impossible

10. Appearance

0 = appropriate dress and grooming
1 = minimally unkempt
2 = poorly groomed, moderately dishevelled, overdressed
3 = disheveled; partly clothed, garish make-up
4 = completely unkempt, decorated, bizarre garb

11. Insight

0 = present, admits illness, agrees with need for treatment
1 = possibly ill
2 = admits behaviour change, but denies illness
3 = admits possible change in behaviour, but denies illness
4 = denies any behaviour change

HAM-D

As with depressive disorders, symptoms of depression for bipolar patients can be scored with the HAM-D test.

The Problem of Diagnosing Bipolar Disorder

The first rule of medicine in dealing with patients is to do no harm. However, this can be difficult when dealing with a dynamically changing psychiatric illness such as BD. This is because misdiagnosis and subsequent inappropriate treatment of patients with BD can worsen their clinical condition and outcome. One of the main reasons for misdiagnosis is that BD patients usually present for the first time as having a depressive disorder and therefore they are treated with antidepressant medications. This occurs in up to 20 % of the BD cases. However, the inappropriate administration of antidepressants could lead to an increased risk of the patients developing mania or side effects. Compounding this problem is the possibility that inappropriate treatment can lead to patients developing life-shortening co-morbidities such as cardiovascular disease or type 2 diabetes mellitus. Taken together, these factors can result in an increased risk of premature death, reduced quality of life and the possibility of the patients developing other complications such as cognitive impairment. For these reasons, the development of accurate diagnostic tests for more accurately distinguishing these disorders in a clinical setting is of major importance.

Among the major psychiatric disorders, BD is unique in that it covers the range of human emotions. This results in a highly complex disease that can be manifested over a wide variety of symptoms. Furthermore, a patient may suffer from unpredictable cycling between the manic and depressive states, or even suffer from the mixed mania/depression syndrome described earlier. As stated above, phase switching is important since treatment of BD is highly dependent on the current mood state of the patient. Those undergoing a manic episode can be treated with mood stabilizers such as lithium, antipsychotics, anxiolytics and anti-epileptics (e.g. lamictal). Patients in a depressive phase are often treated with a combination antidepressants and mood stabilizers, and those in the euthymic state are normally given a maintenance dose of mood stabilizers and/or antipsychotics.

As with all medications used in psychiatry today, these treatments are used to control symptoms and are aimed at allowing the patients to make a functional recovery so that they can re-engage in their lives. Thus, the currently prescribed medications are not likely to target the underlying pathologies and the use of multiple drugs with distinct mechanisms of action is currently the best strategy in order to attempt an effective treatment. Nonetheless, a review of clinical trials on the use of lithium, lamotrigine, valproate and olanzapine as maintenance treatments showed that none of these were capable of treating both mania and depression and only 50 % of BD patients on average appear to show a positive response. Furthermore, the development of drug-related side effects such as metabolic and cardiovascular diseases can pose serious long-term health issues. One example of this is lithium which works over a narrow therapeutic concentration range. This means that there is only a small window between the therapeutic and toxic doses. Therefore careful management of the dosage is needed to avoid these toxicities, which include nausea, resting tremors and blackouts. Also, chronic lithium treatment has been known to cause other undesirable side effects such as irreversible hypothyroidism, insulin resistance and substantial weight gain. Other drugs in use for BD such as valproate and lithium cannot be used in pregnant females because of the high risk of birth defects (this risk is known as teratogenicity). It is plain to see from all of these issues that there is a need for further research to develop more effective and safer treatments for the individuals who suffer from this disease.

Current Tools and Methods in BD Drug Target Discovery

The human genome project led to the possibility of genome-wide association studies (GWAS), which have attempted to unravel the pathophysiology of BD and other psychiatric disorders. A large meta-analysis of genetic associations in BD found polymorphisms in the genes for BDNF, the dopamine D_4 receptor, D-amino acid oxidase activator and tryptophan hydroxylase 1. All of these gene products have roles in neuronal function. However, a different study found that only one gene encoding the L-type voltage dependent calcium channel was linked with BD. This lack of agreement across different studies suggests that BD is not likely to be caused by a defect in a single gene. Another study found a correlation between a BDNF polymorphism and negative symptom scores in BD patients but this was not statistically significant.

How Do We Identify New Drugs for a Complicated Psychiatric Disorder Such as BD?

Animal Models

One of the most widely used approaches for identification of new drugs in psychiatric research is the screening of compounds in preclinical models. These models usually take the form of animals such as rodents and drugs are tested for effects on behavioural readouts. New drugs can be tested with the target in mind, although this requires parallels in the biological pathways between the model and corresponding human disease. However, researchers who have assessed animal models for psychiatric conditions have concluded that the read-outs lack specificity and objectivity, and therefore have problems of reliability and reproducibility. This is most likely because there are no animal models that recapitulate all of the behaviours associated with a complex disorder such as BD. Another problem of the BD preclinical models is the fact that none of them can truly reflect the disease, a term known as translation. For example, there are no models which can translate the oscillation between mania and depression symptoms, and none can be used to distinguish mania from hypomania. In addition, no attempts have been made to model the cognitive deficits of BD in animals. Finally, several BD symptoms involve changes in emotions and moods, and deciding the emotional state of an animal is something that will probably never happen with any certainty.

> *A friend of mine who is also a researcher in this field is fond of saying: "you can't ask an animal how it feels." I have always wanted to respond with: "well, you can ask"*

In response to the poor translation issue of animal models, there has now been a paradigm shift toward the incorporation of more objective read-outs based on biomarker changes. But there is also another problem here. Even if the same molecule is affected in both humans with BD and the respective animal model, there could still be divergent effects in the associated upstream or downstream pathways such that the findings will not be the same in the two species. For example, the role of cortisol in humans in the stress response is carried out by corticosterone in rodents. Therefore, translation of corticosterone measurements from rodents to humans is not likely to lead to meaningful data. Another factor that should be considered is that male animals are used almost exclusively as preclinical models of psychiatric diseases. Thus, translation of the animal biomarker signature back into humans does not always take into account the potential gender differences in the disease or drug response, which can be profound (this topic is covered in Chap. 9).

Cellular Models

The development of novel drugs for BD and other psychiatric illnesses has come to a standstill due to the difficulties of diagnosis and the poor understanding of the affected molecular pathways. In addition, there has been a high failure rate of new drugs resulting from a current focus on molecular changes in animal models that are not necessarily representative of the human disease. As with the other major psychiatric diseases, BD is widely viewed as a disorder of the brain. However, there are numerous links and suggestions that the disease also has a peripheral component. As explained in the chapters for schizophrenia and MDD, this is not surprising as the brain integrates signals from the entire body and responds to these through reciprocal signals in the maintenance of physiology and homeostasis, including the release of bioactive molecules into the circulation. Furthermore, genetic alterations will affect cells in the entire body wherever the gene is expressed, although the specific effects may be different depending on the location. These observations support the idea that it is possible to understand more about brain function by studying peripheral systems. Since taking samples from the brain of living patients is not a viable option, many researchers are now using specific cells from peripheral sources as surrogate models since these can be acquired with minimal discomfort to the subject.

Recent investigations have shown that higher translation of preclinical findings into the clinic can be achieved using biological samples such as blood serum and cells, obtained directly from the patients themselves. As stated throughout this book, peripheral blood contains circulating molecules such as hormones and cytokines, which can be used as molecular readouts of brain function. In addition, blood also contains peripheral blood mononuclear cells (PBMCs) which are essential components of inflammatory and immune responses. Interestingly, PBMCs express most of the functional neurotransmitter and ion channel receptors that are found in the brain. This includes dopamine, serotonin and norepinephrine G-protein coupled receptors, as well as the glutamate, GABA and acetylcholine ion channel receptors. Importantly, they also contain the intracellular signalling pathways which couple receptor binding to the cellular responses. This allows us to use these cells as a potential novel screening tool for drug profiling. This can be achieved using endogenous reporter systems linked to the activation of receptor signalling cascades. For example, a disease signature can be obtained using PBMCs from BD patients and controls after addition of control drugs such as those that stimulate changes in expression of cytokines. Then, novel drugs can be tested on this system to identify new compounds which "normalize" the cytokine levels (Fig. 6.3). This approach has been termed "cytomics" as it permits the multiplex interpretation of cellular responses. The most important advantage to be gained by using cellular models comes from the fact that they are derived directly from human patients with the disease. Therefore, they are more likely to reflect patient-specific, genetically encoded disease alterations, compared with findings in the traditionally used animal models.

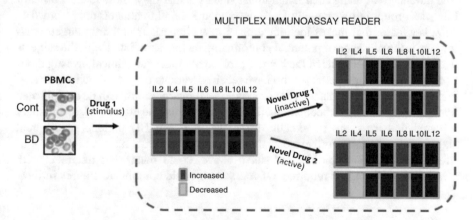

Fig. 6.3 Determining functional aspects of drug effects on cells obtained directly from controls and BD patients using cytomics. Application of a stimulus (Drug1) shows different responses of cytokines in control and BD cells. Testing of an inactive drug (Novel Drug 1) has no effect on this pattern whereas an active drug (Novel Drug 2) reverses the effect so that it is now similar to the control pattern. Therefore this approach can give new insights into the affected pathways in BD and also identify potential targets for development of novel drug treatments

What Are the Key Objectives of BD Research?

Study design is one of the most important aspects for the discovery of new drugs, especially for complex diseases such as BD. Identifying patients before an episode switch from mania to depression or vice versa would be a major leap forward. We already know that there are different potential triggers of episode switching. For example, depressive episodes can be triggered by emotional stressors, such as bereavement or a psychological or physical trauma. Seasonal effects have also been linked with an increased risk of depressive episodes in BD patients. As a potential warning sign, a reduction in the need for sleep often occurs before a manic episode.

Another important point to keep in mind is that most of the biomarker discovery studies carried out to date have simply compared BD patients with healthy controls. In the future, it will be important to include longitudinal data with follow-up information on the development, progression and treatment of the disorder. This is of particular importance in BD studies due to the unpredictable switching of mood states. A single time point study design could also lead to misleading results because of the high rate of misdiagnosis of BD at the disease onset. Finally, one necessity that is often ignored is that of incorporating validation studies by using cohorts from independent clinical sites. The basic rule of science is that everything must be repeated. There are too many one hit wonders out there. There is a clear need for successful repetition of findings multi-centre studies in different cities or even different countries to generate maximum proof that a finding is a valid one. The Netherlands Study for Depression and Anxiety (NESDA) is a good example of such a study as this includes samples from nearly 3,000 patients which have been obtained at three different clinical centres. NESDA also includes follow-up sample collections over several years. The advantage of using such longitudinal studies is that they allow identification of high-risk groups retrospectively, since the outcomes of all treatments are recorded.

A key factor that makes identification of new biomarkers and drug targets challenging is the presence of potential confounding factors associated with the subjects used in these studies. One of the biggest problems is that such clinical investigations normally use chronic patients who have received various drug treatments over several years. Therefore, data analysis and biological interpretation can be complicated by the need to untangle medication effects from true disease effects. Also, one study reported that there are systematic differences between BD patients at different stages of the disease. Furthermore, the fact that BD patients have a higher risk of life-shortening co-morbidities, as stated above, could mean that treatment with other classes of drugs is required. Of course, this could complicate matters further.

What Do Current Drugs Target in BD?

The targets of antipsychotics and antidepressants have been covered in previous chapters. Therefore, this section only describes the known targets of mood stabilizers. The most widely used compound of this class is lithium, which is an inhibitor of an enzyme called glycogen synthase kinase 3 beta (GSK-3β). So how does this work? Mitochondrial GSK-3β normally blocks the activity of pyruvate dehydrogenase, an enzyme involved in conversion of pyruvate to acetyl-CoA. Thus, addition of lithium would remove this inhibition and thereby restore conversion of pyruvate to acetyl-CoA, increasing cellular energy levels in the form of producing more adenosine triphosphate (ATP). This makes sense and it is important as cellular energy levels are thought to be disrupted in BD. However, lithium treatment also appears to affect a number of other pathways via inhibition of GSK-3β. For example, lithium treatment leads to a reduction of programmed cell death, potentially increasing neuronal survival. It also appears to restore normal dopamine levels and normalize cellular signalling pathways such as those involving the phosphoinositide and protein kinase C (PKC) networks and it enhances BDNF activity. The previous chapter on major depression has already described the important role of BDNF in enhancing neuronal function and synaptic connectivity.

Another major drug class used in the treatment of BD is the anti-convulsants, such as valproate. These drugs are also used to treat epilepsy and they appear to work by stabilizing the brain by reducing shifts between the manic and depressive phases, allowing for longer periods of stable function. Previously I described that GABA is the main inhibitory and glutamate the main excitatory neurotransmitter in the brain. Anticonvulsants generally increase the action of GABA and decrease the action of glutamate, thus preventing the high-frequency repetitive firing of neurons as seen in epilepsy. It also appears to explain the anti-mania properties of anticonvulsants like valproate. Animal studies have confirmed that valproate treatment increases synaptic GABA levels by inhibition of GABA degrading enzymes and through blocking re-uptake of the neurotransmitter by neuronal cells. In addition to these effects on neurotransmitter systems, valproate also appears to inhibit histone deacetylase, an enzyme which normally leads to tight packing of DNA in the nucleus of the cell. Such tightly pack regions are generally less active. Thus, inhibition of histone deacetylase by valproate can lead to more transcriptionally active DNA which, in turn, can result in production of important proteins involved in neuronal survival such as BDNF.

The precise therapeutic mechanism of action of these drugs is still under debate. This will be an important issue to address since increased understanding of the pathways involved could lead to identification of novel targets. This is, of course, another area in which biomarkers can help.

Novel Targets in BD

There are many possible new treatment approaches in BD research. Each of these is based on the observed pathways which appear to be disrupted in this disorder (Fig. 6.4). These are described in the following sections.

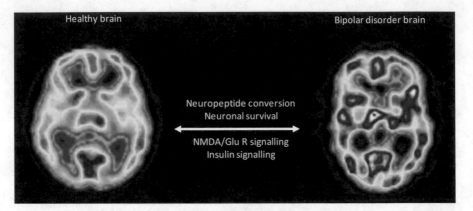

Fig. 6.4 Biological processes affected in bipolar disorder which suggest possible new treatments approaches through the targeting of these pathways. The image shows PET analysis of healthy and bipolar disorder brain in the manic state. The *red*, *yellow* and *blue* colours indicate areas of high to low energy usage and this is clearly higher in multiple areas of the bipolar disorder brain. *NMDA/ Glu R* NMDA subtype of glutamate receptors

Neuropeptide Converting Enzymes

Many neuropeptides and hormones are synthesized initially as larger polypeptide precursors which are then subjected to limited proteolysis by converting enzymes to produce the mature forms. Importantly, the expression levels of many neuropeptides and hormones are known to be altered in psychiatric disorders. For this reason, recent investigations have focused on neuroendocrine precursor processing enzymes such as prolyl endopeptidase (POP). This enzyme is involved in the production of regulatory neuropeptides including arginine-vasopressin and oxytocin, which have been found to be dysregulated in several psychiatric disorders. One study found that measurement of POP activity levels in plasma can be used as a way of differentiating between subjects with untreated mania, schizophrenia and major depression. The study showed that POP activity was increased in schizophrenia and BD patients, but decreased in major depression. Furthermore, treatment of the BD and depressed patients resulted in normalization of the POP activity levels in both cases. However, it should be noted that studies of POP may be a non-starter as it has a widespread tissue distribution and it is altered in other numerous other diseases. In addition, circulating POP activity shows seasonal changes in healthy volunteers and it appears to be affected by changes in diet. For all of these reasons, it is not likely to be a specific biomarker of BD.

Inhibitors of Cell Death (Apoptosis)

Changes in Ca^{2+} flux in BD and other psychiatric disorders are known to induce mitochondrial membrane depolarization, release of cytochrome C and activation of programmed cell death (also known as apoptosis). Apoptosis is actually a form of regulated cell death that does have advantages in multi-celled organisms like humans. For example, during foetal development, apoptosis is responsible for removing tissue once it is no longer needed, as in the separation of the digits by removal of the unwanted flesh in between fingers and toes. However, apoptosis can be a disadvantage if it is triggered unnecessarily in tissues such as the brain. The BD drug valproate is thought to work by increasing the levels of anti-apoptotic proteins in the mitochondria, resulting in increased neuronal survival and enhanced synaptic function. There are also reports of studies using Ca^{2+} channel blockers such as dil-tiazem and verapamil in treatment of BD, although these have had inconsistent results. However, these compounds appear to be more effective in cases of treatment-resistance. Therefore, considerable further work is required in order to unravel the pathways involved in inhibition of apoptosis. This could be fruitful as it could lead to new drug targets in the treatment of BD and other disorders associated with decreased neuronal survival.

NMDA Receptor Antagonists

The NMDA glutamate receptor antagonist ketamine has now been used with some success in treating BD patients during depressive episodes. Amazingly, improve-ments in depressive symptoms and reduced ideas of suicide have been observed within 40 min of giving patients this compound. Treatment of BD patients with ketamine is currently being tested in clinical trials. In BD patients with treatment-resistant depression, ketamine was found to be an effective add-on therapy when co-administered with lithium or valproate. So what effect does ketamine have on the brain? An imaging study showed that ketamine administration altered glucose metabolism in areas of the brain known to be involved in mood disorders, suggest-ing that these alterations may be partially involved in the mechanism of action. Thus, treatment with ketamine may help by reversal of the metabolic dysfunction found in these brain regions.

Insulin Sensitizing Agents

Another possible new target stems from the fact that some patients with BD show signs of insulin resistance or impaired insulin signalling, as described earlier for schizophrenia and depression. Therefore the use of antidiabetic drugs may be a viable approach. Chapter 4 described how treatment of schizophrenia patients with the antidiabetic drugs metformin and rosiglitazone led to improvements in insulin sensitivity and reduced the weight gain associated with administration of antipsy-chotic drugs. In addition, treatment with insulin sensitizing agents may even improve neurological symptoms as found in trials which showed beneficial effects of pioglitazone on cognitive performance of patients with Alzheimer's disease. In a similar manner, a clinical study showed recently that co-administration of intranasal

insulin along with the standard treatments led to a significant improvement in neurocognitive function in euthymic BD patients. For these reasons, further studies are
planned on investigating potential metabolic dysfunctions such as insulin resistance
in early onset BD patients with the overall aim of identifying new targets in this
pathway for early intervention.

Future Prospects in BD Research

The single target approaches used to investigate BD in the past have not led to a
substantial increase in our understanding of the disease aetiology or in the development of new drugs. Therefore, there is a pressing need to interpret existing data
from a different angle. Systems biology approaches are now emerging which
account for multiple targets of different types simultaneously. It is intuitive that that
this is the logical way forward because living organisms function as highly integrated and responsive systems instead of isolated units. The combination of information from the imaging, molecular and cellular profiling techniques described
above and the translation of the findings into the clinic, will lead to deeper insights
into the underlying pathology of complex diseases such as BD and pave the way for
new and improved drug treatments.

Novel therapeutic biomarkers for BD should be researched in order to determine
their ability to discriminate the different phases of this disorder from other mental
disorders that show similar symptom profiles. This may lead to the development of
a biomarker test for correct identification of BD compared with MDD and may also
enable prediction of a phase shift from mania to depression or vice versa. This
would help in preventing subsequent harm of patients resulting from suboptimal or
inappropriate treatment. With this objective in mind, two multiplex immunoassay
profiling studies by Haenisch and co-workers showed that plasma levels of the hormones insulin and C-peptide, and the inflammatory protein matrix metalloproteinase were altered in BD patients in both the mania and depressed mood states. In
contrast, the growth factor-related protein sortilin was changed only in patients who
were in the manic state and the inflammatory factors haptoglobin and chemokine
CC4 appeared to be altered specifically in patients who were in a mixed mood state.
In addition, a study of 69 BD patients and 58 controls showed that serum zinc levels
were significantly lower in patients who were in the depressive phase as compared
with those in the mania phase. Taken together, these findings indicate that it may be
possible to identify both trait- and state-specific biomarkers in BD which could be
useful for helping to predict or monitor mood changes in these patients.

Finally, it would be important to examine the effects of novel treatments such as
antidiabetic agents on dysfunctional pathways seen in BD patients including the
poor metabolism and energy production in the frontal cortex (known as hypofron-
tality). In the initial phases, this will most likely require emerging preclinical drug
testing platforms such as the cytomics approach described earlier in this chapter.
This has the advantage that the cells used would actually be obtained from the

patients themselves and so the results are likely to be translatable to human clinical studies. However, even this is likely to be difficult. Instead of being a single disease, BD appears to be comprised of diverse disease subtypes in addition to having distinct phases. Furthermore, there are well-known differences among the patients in the responses that they show to drug treatment. Therefore, future drug studies should consider the combined use of the traditional clinical rating systems with disease- and state-specific biomarkers as essential for stratification of patients prior to initiation of any clinical studies. This will help to ensure that novel medications are targeted towards the correct population for maximum benefit to the patients.

Chapter 7
The Worrying Case of Anxiety and Stress-Related Disorders

Over the last few decades, evidence has been emerging that many psychiatric disorders such as schizophrenia and depression can involve perturbations of stress pathways throughout the body. Variations in the way that these effects are manifested could be related to the differences in clinical symptoms between affected individuals and in the observed differences in response to treatment. Such differences can also arise from a combination of innate biological perturbations or a precipitating stressful event in the environment. The environmental effects could be either physical stressors such as injury or emotional stresses including loss of a loved one, social defeat or loss of status. Interestingly, there is also emerging evidence that stress during pregnancy can even lead to psychiatric disorders and other illnesses in the offspring many years later. This chapter reviews the effects of such environmental factors on the physiology of the brain and the entire body and provides an overview of the effects that this can have in the precipitation of anxiety-related disorders. There is an unmet clinical need in this area to improve our characterization of the core symptoms of anxiety and to look for novel treatments. This chapter describes what is already known about the physiological pathways involved in anxiety, again involving the whole body concept. It also describes the various scientific methods in place for identifying predisposing factors at the level of gene polymorphisms, as well as those for understanding the molecular and functional changes in the brain and other circuits in the body. Finally, it describes the most commonly used treatments for these disorders and highlight the fact that considerable further improvements are essential.

What Happens When We Get Stressed?

During a stressful event, molecules such as adrenaline and glucocorticoids can be released directly into the bloodstream and this causes a chain reaction of events inside the body. As we have already seen, chronic activation of these pathways can

© Springer International Publishing AG 2017
P.C. Guest, *Biomarkers and Mental Illness*,
DOI 10.1007/978-3-319-46088-8_7

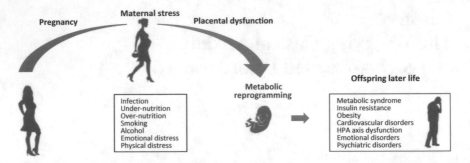

Fig. 7.1 Fetal programming hypothesis. The intrauterine environment is important for development of the fetal organs and tissues. Any deviation from an ideal environment, such as physical or emotional stress, may have long-lasting effects on the offspring

have serious long-term effects on an individual and may even result in development of a psychiatric disorder. The effects of stressful situations experienced during pregnancy have also been investigated and found to have a critical effect on health of the offspring. In the late 1980s, David Barker proposed the "fetal programming hypothesis", which highlights the importance of the intrauterine environment in the development of the fetal organs and tissues. The hypothesis suggests that any deviation from an ideal environment, such as over- or under-nutrition, may have long-lasting effects on organ structure and function. Of course this idea can also be extrapolated to the central nervous system, leading to the possibility that a perturbed intrauterine environment during the first and second trimesters of pregnancy will result in negative effects on the structure of the fetal brain, which may continue on into adulthood (Fig. 7.1). This can result in a variety of health problems in later life of the offspring, such as the development of metabolic syndrome, obesity, cardiovascular diseases and central nervous system disorders.

Correlations between prenatal maternal stress and behavioural and psychological abnormalities in animal offspring have been reported. Such studies have led to the idea that hormones released into the bloodstream in response to a stress stimulus that is either experienced or perceived by the mother, can have a direct effect on development of the brain and other organs in the developing foetus. This is because tight regulation of hormonal release during pregnancy is critical for correct fetal development and any deviation from the normal concentration of these hormones can produce microscopic and macroscopic changes in the brain. This can include alterations in synaptic connectivity within and across distinct brain regions. In fact, the development of behavioural and psychological conditions such as schizophrenia, autism and ADHD appear to be linked to perturbations or hormones in the HPA axis and in other organs of the diffuse neuroendocrine system as described earlier (Fig. 7.2).

Although the current research shows that there is some correlation between increased maternal stress and behavioural and psychological problems in the offspring, this may not be due to hormonal effects alone. The research also suggests

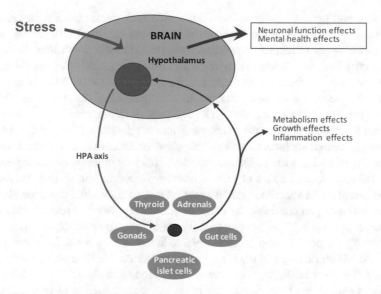

Fig. 7.2 Effect of stress on hormonal circuits of the body resulting in perturbations of whole body health

that genetic links may exist for many of these diseases and they may at least partly result from hereditary causes. It should also be considered that mothers with psychological problems or those who have problems coping with stress may be more likely to become stressed during pregnancy and thus pass these effects on to the offspring. The likely scenario is that these disorders have both hormonal and genetic causes. This is called a "two hit" causation, which suggests that a genetic susceptibility exists in the foetus and then an anomalous hormonal environment in the uterus allows an abnormality to develop.

The remaining sections of this chapter will attempt to summarize what is now known about the effects of stress on the development of the fetal brain and how this may affect the behaviour, psychology and overall health of the offspring later on in life. Since most of the studies to date have used animal models to test these hypotheses, it is only possible to speculate on the extent to which some of these findings can be extrapolated to human psychiatric diseases.

What Is Stress?

Stress as any event, whether real or perceived, that acts to disturb the homeostatic balance in and organism. Most individuals would define stress in everyday life through the effect it has on our emotions. This would include stressors such as looming important deadlines, crucial examinations, overbearing bosses and having to make life-changing decisions. However, not everyone responds to stress in the same way and people appear to have different stress thresholds. Studies aimed at increasing our understanding of underlying causes of these differences will be important in order to identify novel treatments for stressful conditions. Some individuals only become stressed in extreme situations, whereas others could become over stressed in the face of seemingly small matters. Stress can be induced by fear of the unforeseen, the unknown or even an anticipated bad outcome. One example of this is the fear that a loved one will have a serious injury, even though there is no evidence to suggest that this may happen. There are also completely different kinds of stressors which include those of a physical or biochemical nature. Examples include experiencing a high altitude, enduring extremes of temperature, ingestion of a harmful substance or a lack of protein or other important nutrients in the diet.

The extent that a specific stressful event affects an individual can depend simply on that person's perception of the event as something stressful and whether or not they can cope with it. There are two basic types of response called "submission" and "resilience". A person who feels unable to cope with a stressful event can be described as submissive and may respond by experiencing distress or even depression. On the other hand, the same event could lead to resilience in another person and could even boost them to not lose hope, promote a feeling of well being or to work harder at overcoming such problems. The latter condition has been called "eustress" and some people who describe themselves as working well under pressure or having a higher stress threshold may fall into this category.

Diagnostic Tools

DSM-5

The definition of anxiety in DSM-5 is as follows:

A. A persistent fear of one or more social or performance situations in which the person is exposed to unfamiliar people or to possible scrutiny by others. The individual fears that he or she will act in a way (or show anxiety symptoms) that will be embarrassing and humiliating.
B. Exposure to the feared situation almost invariably provokes anxiety, which may take the form of a situationally bound or situationally predisposed panic attack.
C. The person recognizes that this fear is unreasonable or excessive.
D. The feared situations are avoided or else are endured with intense anxiety and distress.
E. The avoidance, anxious anticipation, or distress in the feared social or performance situation(s) interferes significantly with the person's normal routine, occupational or academic functioning, or social activities or relationships, or there is marked distress about having the phobia.
F. The fear, anxiety, or avoidance is persistent, typically lasting 6 or more months.
G. The fear or avoidance is not due to direct physiological effects of a substance such as drugs or medications, or a general medical condition or another psychiatric disorder.

Beck Anxiety Inventory

This is a self administered test in which the subject reads each item in the list and indicates how much they have been bothered by that symptom during the past month (including the day of the test) by circling the number in the corresponding space in the column next to each symptom.

0=not at all			
1=mildly but it didn't bother me much			
2=moderately—it wasn't pleasant at times			
3=severely—it bothered me a lot			
Numbness or tingling	0	1	2	3
Feeling hot	0	1	2	3
Wobbliness in legs	0	1	2	3
Unable to relax	0	1	2	3
Fear of worst happening	0	1	2	3
Dizzy or lightheaded	0	1	2	3
Heart pounding/racing	0	1	2	3
Unsteady	0	1	2	3
Terrified or afraid	0	1	2	3
Nervous	0	1	2	3
Feeling of choking	0	1	2	3
Hands trembling	0	1	2	3
Shaky / unsteady	0	1	2	3
Fear of losing control	0	1	2	3
Difficulty in breathing	0	1	2	3
Fear of dying	0	1	2	3
Scared	0	1	2	3
Indigestion	0	1	2	3
Faint / lightheaded	0	1	2	3
Face flushed	0	1	2	3
Hot/cold sweats	0	1	2	3

A total score of 0–21 indicates very low anxiety. However, it should be noted that too little anxiety could indicate that a person is detached from themselves, or from other people or the environment. Of course, none of these are good scenarios. A score of 22–35 indicates the likely presence of anxiety and a score greater than 36 indicate the likely presence of extreme anxiety and the person should consider consulting a physician or counsellor if the feelings persist.

Stress in Humans

In reaction to a stress-producing stimulus, the body produces a response via activation of the HPA axis. This is a necessary mechanism as it prepares the body for the well-known "fight or flight" reaction to a potentially dangerous situation as described

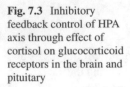

Fig. 7.3 Inhibitory feedback control of HPA axis through effect of cortisol on glucocorticoid receptors in the brain and pituitary

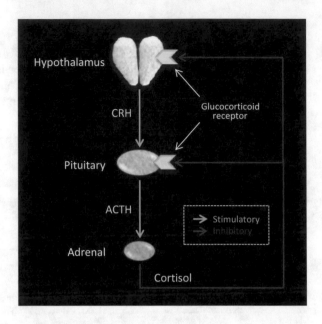

earlier. Interestingly, the stress response produced does not differ greatly depending on the stressor. There are generalized stress responses, which are associated more with the physiological aspect of the stress response. There are also specialized stress responses which are more dependent on the nature of the stress. For example, contraction of the muscles in the arm and pulling a hand away is useful if the hand in question is in contact with a hot surface or freezing cold. However this specialized mechanical type of stress response may not be as appropriate under other conditions.

The main stress response in humans involves activation of the HPA axis via release of CRF, ACTH and cortisol as described in the earlier chapters. The release of these molecules causes a number of physiological changes involved in the fight or flight response such as increased heart rate, constriction of the blood vessels, widening of the pupils and opening of the respiratory passages. In addition, insulin secretion is inhibited and glycogenolysis is stimulated in the muscles and liver. This means that a higher concentration of glucose will be available in the bloodstream to provide the necessary energy in the response. Under normal conditions, the increased blood cortisol levels result in the inhibition of the release of CRF and ACTH in a negative feedback loop (an in-built off switch) (Fig. 7.3).

Foetal Programming in Response to Stress

Prenatal stress has been linked with morphological changes in the fetal brain although the exact mechanisms underlying these effects have not been well elucidated. A number of studies have investigated whether or not the maternal and foetal HPA axes or the placenta are involved in mediating such morphological effects. During pregnancy the mother's body undergoes many physical and chemical changes including altered production of certain hormones. For example, levels of the stress-related hormone cortisol are typically found to be elevated in pregnant mothers. This is important for pregnancy as cortisol is essential for the growth of the fetus and it stimulates the production of surfactant. However, excessively elevated levels of cortisol can actually have negative effects on foetal growth. As explained above, stressful situations can lead to cortisol levels that are too high and this in turn can induce changes in the levels of many other circulating hormones and biomolecules which can also cause changes to the fetal growth environment. Prenatal stress affects the placenta in particular, decreasing the function of this organ via decreased levels of the nutrients and oxygen that normally reach the fetus. This occurs through increased levels of adrenaline in the blood which causes increased vascular resistance and consequential restricted blood flow to the placenta. Another problem linked with stress during pregnancy is the likely concomitant increase in pro-inflammatory cytokines, which has also been linked with neurodevelopmental problems.

The fetus has a natural barrier enzyme to maternal cortisol exposure called 11 β-hydroxysteroid dehydrogenase-2 (11β-HSD2). This molecule normally oxidizes cortisol to produce the inactive metabolite cortisone, thus preventing illicit activation of the stress response. However, the levels of 11β-HSD2 are decreased in instances of maternal stress, leaving the fetus more susceptible to cortisol exposure. The reduction in 11β-HSD2 also leads to an increase in the production of other hormones such as prostaglandins, oestrogens, glucose transporters and placental lactogen. In addition, placental CRF levels are regulated by the maternal HPA axis and research has shown that stress-induced increased concentrations of maternal CRF can lead to structural changes in the hippocampus of the offspring.

Behavioural and Psychological Problems Resulting from Stress

A number of different psychological conditions have been linked to antenatal stress. Studies in rhesus monkeys have demonstrated an increased incidence of ADHD in the offspring when the mothers were exposed to loud noises during pregnancy. Human studies have shown that children of mothers exposed to long-term stressful situations have greater responses to stressors themselves, including abnormalities in coping with stress and exhibiting disturbed social behaviour. The incidence of ADHD was also found to be increased in the offspring of mothers who experienced bereavement during pregnancy. An unexpected bereavement resulted in a 72% greater chance of ADHD occurring in the children. Many other researchers found a higher prevalence of ADHD in the offspring if the mother underwent a serious stressful condition during pregnancy.

Natural disasters can also have effects on offspring health and result in a higher incidence of mental disorders (Fig. 7.4). One study showed that there was a higher

Famine   Hurricane   Earthquake

Fig. 7.4 Natural disasters have been linked to poor health and higher incidence of mental disorders in offspring who were in utero at the time of the disaster. The *left panel* shows a US air force plane dropping food parcels during the Dutch Hunger Winter in 1945. The *middle panel* shows the landfall of hurricane Katrina on August 29, 2005 which devastated the city of New Orleans in the USA. The *right panel* shows a street scene following the largest earthquake in recorded history (magnitude 9.5) which occurred of the coast of Valdiva in Chile on May 22, 1960

incidence of depression in the offspring of mothers who were exposed to a severe earthquake compared to the children of mothers who were not exposed. This is interesting from the point of view that depression has also been linked to abnormalities seen in the HPA axis of pregnant rats exposed to a prenatal stress. After Louisiana was hit by a series of hurricanes and tropical storms, one study found a sevenfold higher increased incidence of autism in the offspring of mothers who were considered to have had high exposure to these weather conditions compared to those in a low exposure group. Also, rat studies have shown that the offspring of mothers who were restrained in the last week of pregnancy are more likely to self administer drugs, such as cocaine and amphetamines, than those from non-restrained mothers. This suggests that there may also be a link between stress and drug dependency.

A number of reports have suggested that there is an increased incidence of schizophrenia in offspring of mothers who have been stressed during pregnancy. Examples include death of the father, as found in a large Finnish study, and even severe weather conditions such as prenatal exposure to a tornado. The German army invasion of the Netherlands in May 1940 during World War II naturally resulted in stress that was nationwide. A study was carried out on a cohort of the offspring of mothers exposed to this stress during the first, second and third trimesters of pregnancy. This revealed a higher incidence of schizophrenia in these subjects, with the largest increase seen in those offspring whose mothers were exposed to this stress during the first trimester. This suggests that the first trimester may be the most critical period for normal neurological growth. Similar effects can also result from another type of stress such as a lack of essential nutrients in the diet. Such a situation again occurred during World War II. During the winter of 1944–1945 when the Nazis destroyed major transport roads and took control of ports in the Netherlands, which led to food shortages for approximately 40,000 individuals. In particular, the protein supply was severely reduced. This led to a greater than twofold increase in the incidence of schizophrenia in the offspring who had been conceived during this period, which is now referred to as the "Dutch Hunger Winter". The largest effects on the prevalence of schizophrenia, as well as metabolic conditions such as diabetes, were seen in subjects who had been conceived during the peak of the famine, when the protein levels were lowest. Similar findings occurred as a result of the Chinese Famine of 1958–1961. Again, the risk of schizophrenia and other conditions was increased by more than twofold in those individuals who had been conceived at the peak of this famine.

The mechanism of how low protein in utero affects mental health in later life could be due to the simple fact that some amino acids are required as precursors (or building blocks) of neurotransmitters. Dopamine, adrenaline and noradrenaline are all synthesized from the amino acid tyrosine and serotonin is synthesized from tryptophan. In cases of famine, such as those described above, it is likely that sources of these amino acids would be depleted and the mother's body would become deficient in proteins, neurotransmitters and other molecules made from that amino acid. As mentioned above, serotonin is a key neurotransmitter involved in HPA axis regulation. Thus, it is possible that a lack of serotonin (via a tryptophan shortage) could be one reason for dysfunction of the HPA axis in mothers exposed to these types of stressors. Furthermore, tryptophan is an essential amino acid which means that it can only be obtained through the diet.

Further evidence supporting the importance of intrauterine environment in the precipitation of psychiatric illnesses comes from animal model studies. Similar to the case of the Dutch Hunger Winter above, studies involving low protein diets in pregnant rats have shown that this can have long-term detrimental effects on the brain in the offspring, including decreased neurotransmitter activity. Other studies have shown that restriction of protein intake during pregnancy may have a negative effect on fetal brain development, which may explain the higher incidence of psychiatric disorders like schizophrenia in the Dutch Hunger Winter and Chinese Famine, outlined above. Interestingly, the behavioural abnormalities seen in these protein restricted models do not even appear until early adulthood. One of these tests is called pre-pulse inhibition which is used as a schizophrenia-like behaviour. In this test, an animal is given a minor stimulus (the pre-pulse) as a warning before a full stimulus is applied. Normal animals will usually display a lower startle response during the second stimulus. However, animals with behavioural abnormalities such as a schizophrenia-like condition will show a startle response of the same magnitude regardless of whether a pre-pulse is given.

Effects of Stress on the Brain

Different regions of the brain are known to be negatively affected by stress at both the macroscopic and microscopic levels. These regions include the hippocampus, amygdala, corpus callosum, cerebral cortex, cerebellum and hypothalamus. The effects have been linked with certain psychological and behavioural problems such as those seen in psychiatric disorders. Also, PET imaging of adult brains has shown increased energy consumption in multiple brain regions of patients diagnosed with

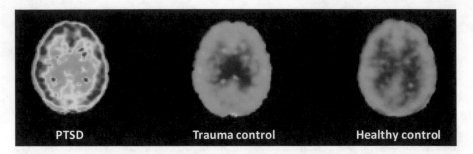

Fig. 7.5 PET imaging of a patient who experienced trauma and developed PTSD (*left*), compared to one who experienced trauma and was not affected (*middle*) and a healthy control who was not exposed to trauma (*right*)

post-traumatic stress disorder (PTSD) after a traumatic incident, compared to those who did not develop PTSD even after experiencing a trauma (Fig. 7.5). As stated earlier in this chapter, the intriguing point here is what makes some people susceptible to developing a behavioural disorder while others are more resistant, even when both are exposed to similar stressors or traumas?

Changes in the size of the corpus callosum and the number of specific cell types within this brain structure have been linked with autism, ADHD and schizophrenia. The hippocampus is involved in the formation of memories and plays a major role in learning and there are reports that prenatal stress has negative effects on memory in rats. Hippocampal granule neurons in the dentate gyrus region continue to be created throughout life and are responsible for the formation of memories. However, the number of such granule neurons was found to be reduced in adult rats with mothers who had been exposed to stress while they were pregnant. Furthermore, this study showed that prenatal stress reduced hippocampal cell proliferation and survival, leading to a reduced number of differentiated neurons. Another interesting finding which came from this study is that the deleterious effects of the prenatal could be counteracted by neonatal handling.

Although animal models have given us some insight into the connection between morphological changes in the brain, behavioural disorders and prenatal stress, the same effects may not necessarily occur in humans. In fact, there are large differences across animal species in the time taken and the extent to which the brain undergoes development before, during and after birth.

Timing and Severity

A study using rats found a 64 % increase in the production of the stress-related hormone corticosterone after handling pregnant rats in the last week of gestation and placing them in unfamiliar cages. Studies aimed at determining the timing and severity of a stressor which affects a human fetus are more difficult. This is mainly due to the obvious fact that it would be unethical to expose pregnant mothers to stress at different points during their pregnancies. However, the occurrence of natural disasters can be used for this purpose in an anecdotal way, as these can provide a large cohort of people affected by the same stress inducer. Such cohorts could be used for retrospective assessment of the offspring of mothers who were pregnant during the disaster.

A study termed "Project Ice Storm" investigated the timing of the effects of maternal stress on the offspring using a large cohort of over three million people who were exposed to extreme cold due to power outages. At the macroscopic level, this investigation found an abnormality in fingerprint formation in offspring from mothers exposed to the cold between weeks 14 and 22 of pregnancy. Since fingerprint formation is known to overlap with development of the hippocampus, follow-up studies set out to determine whether changes in the hippocampal function could be seen in the same individuals. One study found that the offspring of the stressed mother showed cognitive deficits and impaired language abilities indicative of effects on this and other brain regions.

It must be acknowledged that the effects of a particular stressor on a mother depends not only on the nature of the stressor itself but also on the severity of the perceived threat as well as the mother's stress tolerance and behaviour in response to that stressor. Remember that even though individuals can experience the stressful event, not all necessarily respond in the same way.

Potential Advantages of Stress

An interesting and potentially useful aspect of this research is that a region of the hippocampus called the dentate gyrus is more active during acute stress. This is the area of the brain responsible for making new neurons in the processes of neurogenesis and synaptogenesis, which occur in learning and memory formation (Fig. 7.6). In response to a stress situation and consequent HPA axis activation, neural precursor cells in the dentate gyrus are transformed into neurons and glial cells, which are both used in building new synaptic connections. This is a requirement for memory formation since new neuronal networks allow the subjects to remember the object, situation or event that caused the stress. This process appears to be enhanced following emotional recognition of the event by the amygdala region of the brain, which stimulates the hippocampus to form the new connections, presumably so that the individual can attempt to avoid similar situations in the future.

In contrast to the situation with acute stress, chronic stress conditions are thought to decrease the ability of an individual to form new memories. Studies have demonstrated that chronically stressed rats show decreased performance on a spatial maze test compared to non-stressed control rats. Furthermore, studies in humans have shown that several years of living in a stressful situation can lead to increased incidences of depression and other psychiatric conditions. Therefore, one interpretation of these findings is that acute stress may be beneficial and chronic stress could be detrimental to the mental health of an individual.

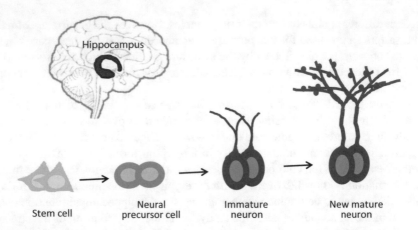

Fig. 7.6 Neurogenesis takes place in the hippocampus and is involved in memory formation. This process is stimulated by acute stress but may be inhibited under chronic stress or traumatic conditions

Effects of Stress on Insulin Resistance and the HPA Axis

In humans, prenatal maternal stress during pregnancy has been linked with a lower birth weight and preterm delivery of the offspring. In addition, stressful conditions during development have been linked with metabolic dysfunctions such as insulin resistance and/or metabolic syndrome in later life. A more recent study tested this by analyzing the levels of glucose and insulin during a glucose tolerance test of young adults whose mothers had experienced stressful life events during their pregnancies. Although there were no significant differences in glucose response, subjects with antenatally stressed mothers had significantly higher insulin levels two hours after glucose consumption compared to the controls. Another study by the same researchers showed that adults whose mothers were stressed during pregnancy had increased cortisol levels in response to the Trier Social Stress Test and they also had a decreased cortisol response after administration of the ACTH stimulation protocol. These studies provide evidence in humans of an association between prenatal stress exposure and alterations tending towards desensitization of insulin signalling and perturbation of HPA axis function in the offspring (Fig. 7.7).

One study showed that prenatal stress induced long-term changes in rats in parameters such as feeding behaviour, glucose metabolism and insulin signalling, as observed in type 2 diabetes mellitus. The researchers who carried out this study suggested that the effect was due to stress-induced increased glucocorticoid levels in utero. Another study tested the effects of administering stress hormones directly into sheep in the early stages of pregnancy and found this caused impaired glucose tolerance and hyperinsulinaemia in the offspring. Again, these findings indicated that glucocorticoid exposure in early pregnancy might lead to long-term metabolic conditions which can also affect neuronal function.

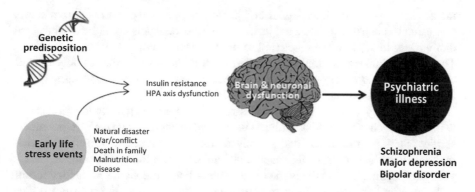

Fig. 7.7 Schematic diagram showing interaction between genetic predisposition and early life environmental factors on neuronal function and later development of psychiatric illnesses

Biomarker Identification for Stress-Related Disorders

A number of studies have been carried out which have shown that patients with anxiety disorders are susceptible to development of other psychiatric conditions. A recent investigation showed that serum biomarkers can be used to predict progression of individuals with social anxiety disorders towards a depressive episode. The researchers carried out a multiplexed immunoassay analysis of serum samples from 72 social anxiety disorder patients recruited as part of the NESDA project over a two year follow-up period. They found that readings of four serum analytes (AXL receptor tyrosine kinase, vascular cell adhesion molecule 1, vitronectin and collagen IV), combined with symptom scores (Inventory of Depressive Symptomatology and Beck Anxiety Inventory somatic subscale), lifetime history of a depressive disorder diagnosis, as well as the BMI values of each subject allowed correct identification of 17 out of the 22 anxiety patients (77 %) who later developed a depressive disorder. Furthermore, these same parameters enabled correct identification of 41 out of the 50 patients (82 %) who did not develop depression. These biomarkers represented diverse molecular classes although most appeared to be involved in the inflammation response, consistent with studies on most other psychiatric disorders. As a point of interest, neither serum biomarkers nor clinical data alone had good performance for predicting depression. This demonstrates the importance of combining biological markers as well as clinical scores and other data in the development of multifaceted algorithms for disease course predictions in psychiatry.

Oxidative damage has been implicated in a variety of psychiatric disorders, including stress-related conditions. For this reason, another research group measured the levels of markers involved in lipid peroxidation in serum from individuals suffering from generalized anxiety disorder, which is one of the most prevalent mental disorders in the general population. The researchers collected blood samples from 40 patients and 40 control subjects and then tested the serum activities of lipid hydroperoxide, paraoxonase and arylesterase. They found that the lipid hydroperoxide activity was significantly higher and paraoxonase activity was lower in patients compared to controls. In addition, lipid hydroperoxide levels could be used to predict anxiety in 93 % of the cases and to predict healthy controls 92 % of the time. Thus, these biomarkers should be retested in individuals with other anxiety-related diseases as well as other psychiatric illnesses for the purposes of establishing reproducibility and disease specificity of the test.

Another study aimed to identify stress-related biomarkers that could predict conversion to psychosis in at-risk individuals who were attending an early intervention service. Stressful life events and perceived stress were assessed for all 39 at-risk individuals and 44 control subjects and all subjects were followed up for at least 1 year after testing. The analytes measured at baseline were serum prolactin, serum C-reactive protein, serum albumin, serum and saliva cortisol, and plasma fibrinogen. At the end of the follow-up period, 10 at-risk subjects had developed a psychiatric disorder and 29 had not. The subjects who had developed a disorder had higher serum levels of prolactin and lower albumin compared to the non-conversion group and to the healthy controls. There were also significant differences in salivary cortisol levels between the groups. Finally, the scientists carrying out the study found that baseline prolactin levels could be used to predict transition to a psychiatric illness whereas albumin levels appeared to have a protective effect. If these stress-related biomarkers could be validated in larger cohort studies, serum biomarker tests could be developed to identify those at-risk individuals who are more likely to develop a psychiatric illness. This could facilitate early intervention strategies and thereby help to decrease the duration of untreated illness and improve patient outcomes.

Therapeutic Implications

The discovery that hyperinsulinaemia might play a role in late onset disorders in the offspring of prenatally stressed mothers, suggests that drugs which improve insulin signalling and glucose handling, may represent a potential novel treatment strategy. This appears to be a common theme in psychiatric disorders although it should not be surprising given the essential role of insulin in the optimum functioning of most cells of the body. In the case of psychiatric disorders, such as schizophrenia, therapeutic strategies that target the underlying metabolic dysfunction could provide an effective alternative to traditional antipsychotic medications. The success of such medications in treating the insulin resistance that seems to come hand in hand with antipsychotic treatment and with the memory cerebral blood flow deficits in Alzheimer's disease have already been described in earlier chapters.

In addition, the adrenal steroid dehydroepiandrosterone (DHEA) has antiglucocorticoid properties that may prove useful regulating high cortisol levels and glucocorticoid action in the brains of psychiatric patients. Studies using DHEA in combination with standard antipsychotic medications in schizophrenia patients found a significant improvement in the negative, depressive and anxiety symptoms of the disease. Also, administration of DHEA was reported to reduce the extrapyramidal side-effects of some antipsychotic drug treatments such as the involuntary tremors that can sometimes occur.

Conclusions and Future Prospects

This chapter described the remarkable finding that prenatal stress can led to development of psychological and behavioural problems, as well as diseases such as schizophrenia, autism and depression in the offspring in their later lives. Changes are seen in the brains of a variety of animal models in response to a prenatal stress including regions such as the hippocampus, amygdala, corpus callosum, cerebral cortex, cerebellum and hypothalamus. These areas of the brain are responsible for the control of behaviour and their alteration in the disease state could explain the associated psychological problems, although this has not been proven to occur in humans. The mechanisms by which these deleterious changes occur in the brain most likely surround the maternal and fetal HPA axes and the effect of the intrauterine environment on development of the brain. Regardless of any environmental effects on the fetus, it is likely that there is also a genetic element to the development of such conditions. It also appears that the timing and severity of the stress experienced by mothers is an important factor. Finally, there is increasing evidence of metabolic and hormonal abnormalities in conditions such as schizophrenia, which in some cases may be associated with prenatal stress. Abnormalities in the metabolism of glucose, insulin signalling and the HPA axis appear to be present in the early stages of these disorders. The good news is that this may provide the basis for the development of much-needed biomarkers for better detection and categorization of psychiatric conditions. The incorporation of such biomarkers into the clinical setting could lead to improved diagnosis and personalized medicine strategies, and confer the opportunity of potential preemptive treatments. Given the potential of this line of research to improve diagnosis and create alternative treatment strategies, more research is warranted.

It is clear that the environmental triggering of stress in utero is not the only catalyst for development of psychiatric disease. It is likely that there is also a genetic predisposition that when combined with an environmental trigger leads to disease onset or progression. This is in accordance with the two hit hypothesis. All of us have the tendency to dismiss our "emotional" responses like sadness or stress since they are not tangible. However, everything that we feel is due to surges and ebbing of certain chemicals, neurotransmitters and hormones within our bodies which produce the sensations of stress but these also lead to a cascade of other reactions which can have damaging outcomes on physical and mental health.

Chapter 8
The Autism Spectrum Conditions and the Extreme Male Brain Syndrome

This chapter summarizes the current understanding of autism and autism-related syndromes. It explains the process of diagnosis and covers some of the options for education and intervention to help improve the lives of those individuals who suffer from these conditions. Autism can affect young children as well as adults and is often debilitating through disruption of socialization, mood states and intelligence. Paradoxically, in some cases, autism can also result in enhanced abilities in specific areas such as pattern recognition, art, music and mathematics. Furthermore, some people with autism can have normal or even high IQ levels. The chapter also outlines prevailing theories which go some way towards explaining the main symptoms of autism conditions, such as the Extreme Male Brain and the Empathizing-Systemizing hypotheses. In addition, a number of biomarker-related studies have now been carried out which have helped to increase our understanding of the underlying physiological pathways that are altered in individuals with these conditions. Such studies have looked both in the brain and the periphery, in line with the whole body concept of psychiatric conditions. This includes the investigation of predisposing genes, which can increase the risk of developing autism, and environmental factors that could either act separately or in concert with the genetic factors to bring about these conditions. Further advances along these avenues could result in identification of novel biomarkers which in turn could lead to earlier detection and more effective treatments. Although there are still no drugs for treatment of the autism spectrum conditions, it is anticipated that further studies driven by identification and application of physiological biomarkers will soon turn this deficit around. This can only be achieved by increasing our knowledge on the affected pathways in the brain and periphery, and identification of much-needed novel drug targets.

© Springer International Publishing AG 2017
P.C. Guest, *Biomarkers and Mental Illness*,
DOI 10.1007/978-3-319-46088-8_8

What Is Autism?

Imagine that people seem illogical in what they do and social rituals are confusing with their syntax and speech patterns. Imagine that you are constantly aware of electric humming noises when you enter a store and you are struck by sound and light overwhelming your senses. You hide in your books, such as textbooks and these relax you and take you away from the confusing sensory overload. Imagine that you have to constantly guess at the meaning behind peoples' tones and mannerisms and you have no idea about what they are feeling. You feel like everyone around is speaking a foreign language that you don't understand and you feel that they patronize you. Imagine that you find socializing to be scary and unpredictable. You find it difficult to figure out what people expect from you and you always need highly specific instructions. Imagine that you don't like making eye contact because you can't understand the facial expressions that people are making.

Imagine that you can memorize some kinds of information almost instantly. You could entertain yourself for hours by replaying a piece of music in your mind that you first heard when you were 5 years-old. You enjoy patterns when everything repeats itself. You tend to focus and fixate on details to the extent that you can sometimes learn things much faster than others. You find subjects like physics, mathematics and chemistry easier than interacting people. Imagine that you look at a scene for just a few seconds and then draw a picture of this reproducing all of the details. Imagine you can listen to a piece of music just once and then play this back on your piano note for note. You like watching movies over and over again because they are predictable and different from actual life.

These are just some of the features that a person with autism or high functioning autism may experience (Fig. 8.1).

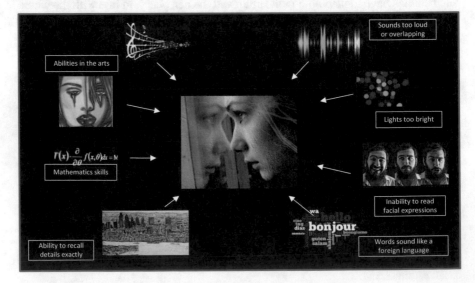

Fig. 8.1 Possible symptoms of autism or high-functioning autism cases

Just to categorize things, autism spectrum disorders (ASDs) are a range of clinically heterogeneous neurodevelopmental conditions consisting mainly of autistic disorder, Asperger's syndrome and pervasive developmental disorder not otherwise specified (PDD-NOS). In addition, there are other forms of ASD that include childhood disintegrative disorder (CDD) and genetic conditions such as Rett's syndrome, fragile X syndrome and tuberous sclerosis. ASDs are characterized by impaired social interaction and communication, along with self-limited and repetitive behaviours. As a complicating factor, individuals with ASD often have comorbidities such as epilepsy, gastrointestinal obstructions, motor deficits, sleep disturbances, cognitive deficits, mental retardation (IQ <70) or ADHD. In the last two decades, the number of reported ASD cases has increased from 1 in 2000–2500 to 1 in 88. This apparent increase could be due to environmental factors, a broadening of diagnostic criteria, as well as increased awareness of the disorder, improved diagnostic methods, or any of these in combination. In most ASD subtypes, a bias in gender occurrence has been well-documented, with an increased prevalence in males over females at a ratio of 4:1, or even higher. Thus, potential perturbations in the sex hormones were the first likely suspects. One hypothesis that takes into account these gender differences and the cognitive manifestations of ASD is the Extreme Male Brain theory. This theory describes ASD as an extreme variation of male intelligence and is based on ground-breaking research on sex differences in cognition and behaviour (see the next chapter for more on this). A number of findings support this theory, including studies showing that the systemizing quotient (SQ—the ability to construct, see and analyze systems) is higher in "normal" males compared to "normal" females, and both males and females with ASD show higher SQ scores than males who do not have ASD. Take the test outlined in Table 8.1 and see how you do.

In addition, the empathy quotient (EQ—the ability to detect feelings in others and respond appropriately) is higher in "normal" females compared to normal males, and both males and females with ASD score even lower than normal males and females. There is a test for this as well which you take by answering the questions outlined in Table 8.2.

Diagnostic Tools

Parents will usually notice the signs of autism which can develop gradually during the first 2–3 years of their child's life. These signs can include the inability of the child to reach certain developmental milestones at a normal pace with other children of the same age or a regression to a less developed state could occur. An actual diagnosis of ASD requires that the signs and symptoms are apparent early in the child's life and this is normally before the age of 3 years. Methods of identifying individuals with ASDs include the Autism Diagnostic Observation Schedule (ADOS) and the Revised Autism Diagnostic Interview (ADI-R), which are both based on the DSM criteria as indicated below.

Table 8.1 Systemizing Quotient (SQ) test. There are no right or wrong answers, or trick questions. Answer with either (**A**) strongly agree, (**B**) slightly agree, (**C**) slightly disagree or (**D**) strongly disagree. Or take the test on line (http://personality-testing.info/tests/EQSQ.php)

1. When I listen to a piece of music, I always notice the way it's structured

2. I adhere to common superstitions

3. I often make resolutions, but find it hard to stick to them

4. I prefer to read non-fiction than fiction

5. If I were buying a car, I would want to obtain specific information about its engine capacity

6. When I look at a painting, I do not usually think about the technique involved in making it

7. If there was a problem with the electrical wiring in my home, I'd be able to fix it myself

8. When I have a dream, I find it difficult to remember precise details about the dream the next day

9. When I watch a film, I prefer to be with a group of friends, rather than alone

10. I am interested in learning about different religions

11. I rarely read articles or webpages about new technology

12. I do not enjoy games that involve a high degree of strategy

13. I am fascinated by how machines work

14. I make it a point of listening to the news each morning

15. In maths, I am intrigued by the rules and patterns governing numbers

16. I am bad about keeping in touch with old friends

17. When I am relating a story, I often leave out details and just give the gist of what happened

18. I find it difficult to understand instruction manuals for putting appliances together

19. When I look at an animal, I like to know the precise species it belongs to

20. If I were buying a computer, I would want to know exact details about its hard drive capacity and processor speed

21. I enjoy participating in sport

22. I try to avoid doing household chores if I can

23. When I cook, I do not think about exactly how different methods and ingredients contribute to the final product

24. I find it difficult to read and understand maps

25. If I had a collection (e.g. CDs, coins, stamps), it would be highly organized

26. When I look at a piece of furniture, I do not notice the details of how it was constructed

27. The idea of engaging in "risk-taking" activities appeals to me

28. When I learn about historical events, I do not focus on exact dates

29. When I read the newspaper, I am drawn to tables of information, such as football league scores or stock market indices

30. When I learn a language, I become intrigued by its grammatical rules

31. I find it difficult to learn my way around a new city

32. I do not tend to watch science documentaries on television or read articles about science and nature

33. If I were buying a stereo, I would want to know about its precise technical features

34. I find it easy to grasp exactly how odds work in betting

(continued)

Table 8.1 (continued)

35. I am not very meticulous when I carry out DIY
36. I find it easy to carry on a conversation with someone I've just met
37. When I look at a building, I am curious about the precise way it was constructed
38. When an election is being held, I am not interested in the results for each constituency
39. When I lend someone money, I expect them to pay me back exactly what they owe me
40. I find it difficult to understand information the bank sends me on different investment and saving systems
41. When travelling by train, I often wonder exactly how the rail networks are coordinated
42. When I buy a new appliance, I do not read the instruction manual very thoroughly
43. If I were buying a camera, I would not look carefully into the quality of the lens
44. When I read something, I always notice whether it is grammatically correct
45. When I hear the weather forecast, I am not very interested in the meteorological patterns
46. I often wonder what it would be like to be someone else
47. I find it difficult to do two things at once
48. When I look at a mountain, I think about how precisely it was formed
49. I can easily visualize how the motorways in my region link up
50. When I'm in a restaurant, I often have a hard time deciding what to order
51. When I'm in a plane, I do not think about the aerodynamics
52. I often forget the precise details of conversations I've had
53. When I am walking in the country, I am curious about how the various kinds of trees differ
54. After meeting someone just once or twice, I find it difficult to remember precisely what they look like
55. I am interested in knowing the path a river takes from its source to the sea
56. I do not read legal documents very carefully
57. I am not interested in understanding how wireless communication works
58. I am curious about life on other planets
59. When I travel, I like to learn specific details about the culture of the place I am visiting
60. I do not care to know the names of the plants I see

How to work out your SQ score

Score two points for each of the following items if you answered (**A**) or one point if you answered (**B**): 1, 4, 5, 7, 13, 15, 19, 20, 25, 29, 30, 33, 34, 37, 41, 44, 48, 49, 53, 55

Score two points for each of the following items if you answered (**D**) or one point if you answered (**C**): 6, 11, 12, 18, 23, 24, 26, 28, 31, 32, 35, 38, 40, 42, 43, 45, 51, 56, 57, 60

The remaining questions are not scored

What your score means

On average women score about 24 and men score about 30

0–19 = a lower than average ability for analyzing and exploring a system

20–39 = an average ability for analyzing and exploring a system

40–50 = an above average ability for analyzing and exploring a system

51–80 = a very high ability for analyzing and exploring a system. Three times as many people with Asperger syndrome score in this range, compared to typical men, and almost no women score this high

Table 8.2 Empathy Quotient (EQ) test. There are no right or wrong answers, or trick questions. Answer with either (**A**) strongly agree, (**B**) slightly agree, (**C**) slightly disagree or (**D**) strongly disagree. Or take the test on line (http://personality-testing.info/tests/EQSQ.php)

1. I can easily tell if someone else wants to enter a conversation

2. I prefer animals to humans

3. I try to keep up with the current trends and fashions

4. I find it difficult to explain to others things that I understand easily, when they don't understand it the first time

5. I dream most nights

6. I really enjoy caring for other people

7. I try to solve my own problems rather than discussing them with others

8. I find it hard to know what to do in a social situation

9. I am at my best first thing in the morning

10. People often tell me that I went too far in driving my point home in a discussion

11. It doesn't bother me too much if I am late meeting a friend

12. Friendships and relationships are just too difficult, so I tend not to bother with them

13. I would never break a law, no matter how minor

14. I often find it difficult to judge if something is rude or polite

15. In a conversation, I tend to focus on my own thoughts rather than on what my listener might be thinking

16. I prefer practical jokes to verbal humour

17. I live life for today rather than the future

18. When I was a child, I enjoyed cutting up worms to see what would happen

19. I can pick up quickly if someone says one thing but means another

20. I tend to have very strong opinions about morality

21. It is hard for me to see why some things upset people so much

22. I find it easy to put myself in somebody else's shoes

23. I think that good manners are the most important thing a parent can teach their child

24. I like to do things on the spur of the moment

25. I am good at predicting how someone will feel

26. I am quick to spot when someone in a group is feeling awkward or uncomfortable

27. If I say something that someone else is offended by, I think that that's their problem, not mine

28. If anyone asked me if I liked their haircut, I would reply truthfully, even if I didn't like it

29. I can't always see why someone should have felt offended by a remark

30. People often tell me that I am very unpredictable

31. I enjoy being the centre of attention at any social gathering

32. Seeing people cry doesn't really upset me

33. I enjoy having discussions about politics

34. I am very blunt, which some people take to be rudeness, even though this is unintentional

35. I don't find social situations confusing

(continued)

Table 8.2 (continued)

36. Other people tell me I am good at understanding how they are feeling and what they are thinking
37. When I talk to people, I tend to talk about their experiences rather than my own
38. It upsets me to see an animal in pain
39. I am able to make decisions without being influenced by people's feelings
40. I can't relax until I have done everything I had planned to do that day
41. I can easily tell if someone else is interested or bored with what I am saying
42. I get upset if I see people suffering on news programmes
43. Friends usually talk to me about their problems as they say that I am very understanding
44. I can sense if I am intruding, even if the other person doesn't tell me
45. I often start new hobbies, but quickly become bored with them and move on to something else
46. People sometimes tell me that I have gone too far with teasing
47. I would be too nervous to go on a big rollercoaster
48. Other people often say that I am insensitive, though I don't always see why
49. If I see a stranger in a group, I think that it is up to them to make an effort to join in
50. I usually stay emotionally detached when watching a film
51. I like to be very organized in day-to-day life and often makes lists of the chores I have to do
52. I can tune into how someone else feels rapidly and intuitively
53. I don't like to take risks
54. I can easily work out what another person might want to talk about
55. I can tell if someone is masking their true emotion
56. Before making a decision, I always weigh up the pros and cons
57. I don't consciously work out the rules of social situations
58. I am good at predicting what someone will do
59. I tend to get emotionally involved with a friend's problems
60. I can usually appreciate the other person's viewpoint, even if I don't agree with it

How to work out your EQ score

Score two points for each of the following items if you answered (**A**) or one point if you answered (**B**): 1, 6, 19, 22, 25, 26, 35, 36, 37, 38, 41, 42, 43, 44, 52, 54, 55, 57, 58, 59, 60
Score two points for each of the following items if you answered (**C**) or one point if you answered (**D**): 4, 8, 10, 11, 12, 14, 15,18, 21, 27, 28, 29, 32, 34, 39, 46, 48, 49, 50
The remaining questions are not scored

What your score means

On average, most women score about 47 and most men about 42. Most people with Asperger syndrome or high-functioning autism score about 20
0–32 = lower than average ability for understanding how other people feel and responding appropriately
33–52 = average ability for understanding how other people feel and responding appropriately
53–63 = above average ability for understanding how other people feel and responding appropriately
64–80 = very high ability for understanding how other people feel and responding appropriately

1. Persistent deficits in social communication and social interaction across multiple contexts, as manifested by the following current or historical examples:

   (a) Deficits in social-emotional reciprocity, ranging from abnormal social approach and failure of normal back-and-forth conversation, to reduced sharing of interests, emotions, or failure to initiate or respond to social interactions.
   (b) Deficits in nonverbal communication in social interactions, ranging from poorly integrated verbal and nonverbal communication to abnormalities in eye contact and body language or deficits in understanding and use of gestures with a lack of facial expressions and nonverbal communication.
   (c) Deficits in developing, maintaining and understanding relationships. This can range from difficulties adjusting behaviour to suit social context to difficulties in sharing imaginative play or in making friends to absence of interest in peers.

2. Restricted, repetitive patterns of behaviour, interests or activities, as manifested by at least two of the following current or historical examples:

   (a) Stereotyped or repetitive motor movements, use of objects, or speech such as lining up toys or flipping objects and repeating or using idiosyncratic phrases.
   (b) Insistence on sameness, inflexible adherence to routines or ritualized patterns or verbal and nonverbal behaviour, such as exhibiting distress at small changes, difficulties with transitions, rigid thinking patterns and greeting rituals, of the need to take same route or eat the same food every day.
   (c) Highly restricted, fixated interests that are abnormal in intensity or focus, such as a strong attachment to or preoccupation with unusual objects, or strongly persevering interests.
   (d) Hyper- or hypo-reactivity to sensory input or unusual interests in sensory aspects of the environment, such as apparent indifference to pain or temperature, adverse responses to specific sounds or textures, excessive smelling or touching of objects, or excessive fascination with lights or movements.

3. Symptoms must be present in the early developmental period but may not become fully manifest until social demands exceed limited capacities, or may be masked by learned strategies in later life.
4. Symptoms cause clinically significant impairments in social, occupational or other important areas of current functioning.
5. The disturbances are not better explained by intellectual disability or a developmental delay. Intellectual disability and autism spectrum disorder frequently co-occur. For a diagnosis of co-morbid ASD and intellectual disability, social communication should be below that expected for general developmental level.

Individuals with an established DSM diagnosis of autistic disorder, Asperger's syndrome or PDD-NOS should be given the diagnosis of ASD and it should be specified whether this occurs with or without intellectual impairment, language impairment, a known medical or genetic condition or environmental factor, or another neurodevelopmental, mental or behavioural disorder.

Idiopathic ASD

In addition to the above, diagnosis may be able to distinguish individuals who have idiopathic ASD from those with symptomatic ASD. In the case of idiopathic ASD, no specific aetiology occurs as in the cases of genetic or neurological disorders linked to ASD (e.g. Rett's syndrome, fragile X syndrome and tuberous sclerosis). Individuals with idiopathic ASD often show abnormal behaviour although their cognitive function is more likely to be normal as in high functioning autism and Asperger's syndrome cases. Although the underlying causes of idiopathic ASD have not been completely explained, it is possible that genetic susceptibility and environmental triggers may be involved.

Symptomatic ASD

In contrast with the idiopathic form of ASD, symptomatic ASD is often associated with mental retardation and abnormal neurological function. Genetic analyses have demonstrated that a greater number of genes are associated with symptomatic ASD cases that have severe co-morbid mental retardation, compared to the idiopathic forms of ASD. However, diagnosis can be complicated by the heterogeneity, type and severity of symptoms. This is because the clinical forms of ASD can share common pathways and result in phenotype overlaps. Due to these problems in defining the different forms of ASD, most of the biomarker studies carried out to date have not even attempted to disentangle the matter of which genes or proteins are associated with specific ASD subtypes or co-morbidities.

What Causes Autism?

A frequently reported physical feature of ASD is a large head size (macrocephaly), with an incidence of 80 % between birth and 2 years of age (Fig. 8.2). Research has shown that this may be caused by an increase in anabolic growth pathways in the central nervous system, as evidenced by transiently increased levels of growth factors such as BDNF. Moreover, some children with ASD have impaired cerebral blood circulation, which can result in hypoxia (reduced oxygen supply) in specifc brain areas. As with other psychiatric disorders, abnormal brain connectivity is likely to contribute to the pathology of ASD by altering the flow of information within and across distinct brain regions. In the case of autism, this appears to be due hyperconnectivity (Fig. 8.3). This may explain the sensory overload that is sometimes experienced by autism patients.

At the genetic level, ASD is hereditable and potential risk genes have been identified in 20–25 % of the cases by GWAS and other types of genetic linkage studies.

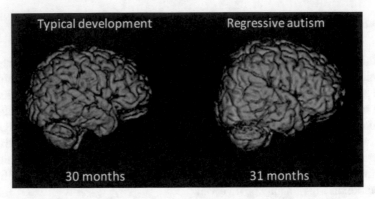

Fig. 8.2 Research has shown that children with regressive autism generally have larger brains than children without the disorder, which appears to be due to increased levels of growth factors like BDNF

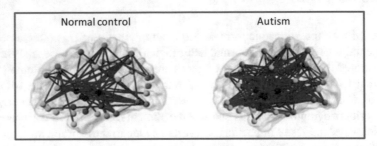

Fig. 8.3 Functional magnetic resonance imaging (fMRI) studies have found significant overconnectivity in the brains of teenagers with autism

The most significant progress in the identification of genetic causes in ASD has come from identifying mutations and by the finding of potential associations with other disorders that predispose individuals towards development of ASD. In addition, abnormalities have been found to occur on virtually all chromosomes in around 5 % of ASD children, as shown by chromosomal analysis or fluorescence in situ hybridization (FISH) studies, which can be used to identify the affected chromosomal regions. The most frequently occurring chromosomal aberrations observed in ASD include a duplication within chromosome 15 (15q11-13) in mothers of ASD children and a deletion or duplication within chromosome 16 (16p11.2), and these each occur in around 1 % of ASD cases (Fig. 8.4). Perhaps not surprisingly, these chromosomal regions contain genes that have been linked to cognitive functions. However, despite these findings, none of the individual or combined risk genes for ASD can be used to predict the extent to which ASD symptoms are manifested (a situation that is common to all psychiatric disorders). As with most diseases, complex gene–environment interactions are thought to be involved in the aetiology of ASD. As examples, exposure to viruses such as rubella during the first trimester of pregnancy, as well as paternal or maternal age can significantly increase the risk of developing ASD. The major functional pathways that appear to be affected in

Fig. 8.4 Schematic
diagram showing the case
of a deletion (*left*) or
duplication (*right*) in the
long arm of chromosome
near the centromere

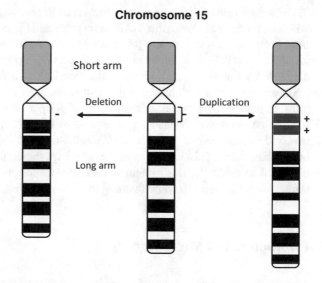

ASD are immune dysregulation, hormonal perturbations, oxidative stress and mitochondrial dysfunction.

How Is Autism Treated?

It is important to note that the existing treatments for ASD do not alleviate core deficits, but instead are designed to treat the observed behavioural symptoms. At present, the only FDA-approved drug treatments for ASD are the antipsychotics
risperidone and aripiprazole, which are used in the treatment of other neuropsychi-
atric disorders like schizophrenia and bipolar disorder. In order to increase the accuracy of diagnoses and develop better ASD treatments, there is now a critical need
for the development of objective diagnostic methods and facilitated by better characterization through the use of biomarkers. Of course, biomarkers could also be
used to identify potential novel drug targets, as described in earlier chapters. Some
of these potential new treatment approaches are described in a few sections.

Biomarkers for Autism

Transcriptomics

Since there is a lack of convincing evidence of gene defects alone causing ASD,
researchers have recently begun to address the question by using molecular profiling
approaches. A cDNA microarray study of ASD brains published in 2001 found that
hydroxy-5-methyl-4-isoxazolepropionic acid (AMPA) receptors may be disrupted

in the cerebellum of autism patients, compared to the findings in control brains. In addition, it has been shown that GABA receptors and the cell adhesion protein reelin have been the most commonly found transcriptomic biomarkers. Given the role of all of these proteins in synaptic plasticity, these findings support the abnormal brain connectivity theory of ASD. Many of the remaining mRNA transcripts affected in ASD are associated with inflammation pathways such as the interferon-related proteins. In addition, transcriptomic profiling of peripheral tissues from ASD patients has shown decreased levels of protein kinase C epsilon, a protein known to be involved in anxiety-like behaviours via modulation of GABA receptor responses to benzodiazepines. This lends support to the hypothesis that the GABA-system is disrupted in ASD. As this system is normally inhibitory, this could explain why sensory input appears to come flooding in all at once in some autism patients.

Proteomics and Metabolomics

To date, there have only been a few studies which have investigated changes in protein and metabolite levels in ASD patients. Proteomic profiling studies have been carried out using serum, plasma and *post-mortem* brain samples using mainly mass spectrometry or multiplex immunoassay profiling approaches. Only a few proteins have been identified that appear to be altered in ASD, but these include the energy regulator creatine kinase, the transport protein ferritin and the cytokine IL-8. Altered levels of eotaxin and CCL5 have also been reported, which are both associated with inflammation. However, sporadic reports have suggested that changes may also occur in other proteins such as BDNF, glial fibrillary acid protein and reelin, which are involved in neuronal development and synaptic connectivity. One study showed female-specific differences in adults with Aperger syndrome in the circulating levels of other hormones and growth factors, such as growth hormone and insulin-related molecules, along with male-specifc changes in inflammatory factors like cytokines and interleukins. Abnormalities in the immune system have been observed previously in ASD, which may be indicative of increased astrocyte activity and inflammation in the brain. In addition, autoantibodies have been identified in children with ASD that are not altered in "normal" control children. This includes antibodies against myelin basic protein, brain extracts, purkinje cells, growth factors and neurofilament proteins. Researchers have suggested that in ASD premature birth and susceptibility genes stimulate the release of pro-inflammatory and neurotoxic molecules, which contribute to brain inflammation and knock-on changes. Therefore, it may come as no surprise that clinical studies have tested the use of anti-inflammatory agents to improve ASD symptoms. Taken together, these findings are consistent with the idea that inflammation and neuronal development are the most commonly affected pathways in ASD and these may be linked.

A possible explanation for the altered synaptic connectivity seen in autism was provided by a recent proteomic profiling study of *post-mortem* brain samples from autism patients and controls. This study focussed on the prefrontal cortex and cerebellum since previous studies have found that these two brain regions are structurally and functionally connected and both have already been implicated in ASD. They used a technique called selected reaction monitoring (SRM) mass spectrometry, which allows the precise measurement of specific molecules of known mass (as opposed to the non-targeted profiling approach described in Chap. 3). Using this method the authors found decreased levels of the immature astrocyte marker vimentin in both brain regions, suggesting a decrease in astrocyte precursor cells. Also, decreased levels of proteins involved in myelination and increased amounts of synaptic and energy-related proteins were found in the prefrontal cortex, and opposite directional changes were found for these same proteins in the cerebellum. Taken together, these findings indicated that ASD was typified by opposite changes on proteins that drive synaptic connections in these two brain regions.

Changes in the metabolites and small molecules found in ASD include the neurotransmitters glutamate, GABA, dopamine, norepinephrine and serotonin, the sex hormone testosterone and the antioxidant molecule glutathione. The changes in the neurotransmitters are intriguing given their role in synaptic function but it is still not known whether or not these are a cause or consequence of ASD. Nevertheless, there might be some value in targeting these pathways as potential therapeutic approaches. Early exposure to high levels of androgenic hormones like testosterone can affect development of multiple tissues of the body including the brain. It is now known that fetal testosterone levels can have permanent effects that predict later development of sex differences in the brain and behavioural patterns. One of these effects is a reduction in empathy levels in later life. Baron-Cohen and colleagues found that increased levels of testosterone occurred in female patients only. The finding of disruptions in antioxidant molecules such as glutathione may also be important and have sparked a recent interest in targeting this molecule as a potential novel treatment for ASD. These results support the theories on altered synaptic functions in ASD and suggest that this may in part be regulated by alterations in the sex hormones and reduction-oxidation (REDOX) networks. However, many of the studies described above have produced inconsistent findings, which is likely to be due to the fact that the patients may have been poorly characterized prior to the study. In fact, very few studies have attempted to match subjects for variables such as the idiopathic or symptomatic forms or even more basic parameters such as age and gender. For such reasons, progress in ASD research has been relatively poor, compared to the study of other neurological diseases. However, the findings described above suggest that further biomarker-based research in ASD guided by rigid patient selection criteria could help to overcome this deficit.

Developing New Treatments for Autism

As stated earlier in this chapter, several drugs are in current use for treatment of autism although all of these are targeted towards management of symptoms or co-morbidities. The drugs used for symptom management include antipsychotics such as risperidone and aripiprazole, antidepressants like fluoxetine, citalopram and clomipramine, mood stabilizers including divalproex and treatments for ADHD, such as methylphenidate, atomoxetine and clonidine. A number of drugs have now shown promise in potentially treating a least some of the underlying symptoms of ASD. However, it should be noted that it is still too early to pick a favourite, if any, as none of these come without their problems. Thus, considerable further research is needed.

Melatonin

This is a neurohormone secreted by the pineal gland, involved in sleep regulation. Clinical studies of ASD children and adolescents treated with melatonin resulted in an improvement in sleep in most of the cases compared children who received placebo. However, melatonin can have side effects including dizziness, nervousness, stomach pain, rash or itchiness.

Omega-3 Fatty Acids

Despite the initial promise, several studies have now reported that supplementation of low levels of omega-3 fatty acids to children with ASD had no effect on behaviours such as social interaction, communication, stereotypy or hyperactivity. Nevertheless, larger trials are still ongoing.

Glutamate and GABA

Considering that glutamate and GABA signalling may be hyper-activated in ASD patients, recent clinical studies have begun testing drugs which antagonize GABA pathways in children and adolescents. Treatment with these compounds had several benefits such as improvements in socializing and attentive behaviours. However, some children developed adverse effects including sedation, irritability and increased seizure frequency.

Oxytocin

Early studies tested the effects of oxytocin infusion in ASD and found that this can reduce repetitive behaviours and improve speech comprehension. More recent studies found that intranasal administration of oxytocin to ASD subjects led to stronger personal interactions, increased eye gaze, improved scores in communication and social interaction tests and enhanced response to facial stimuli. However, the side effects of oxytocin treatment can be considerable such as induction of uterine contractions and lactation in females and it can sometimes cause antisocial behaviour.

The Future of Autism Research

Unlike most psychiatric conditions, ASD is special since there are known treatments available which can be used for improvement of the core symptoms or behaviours. This is mainly because we still don't know what causes ASD and we are likely to be making a critical error by lumping all of these conditions together under the heading "ASD". This shows a lack of foresight as we already know that they are really heterogeneous conditions and are likely to result from different causes based on the differences in behavioural testing as well as the biomarker profiles. Remember that the objective of personalized medicine is to ensure that only the right patients receive the right drug treatments. In the case of ASD, it is clear that the combination of behavioural testing and biomarker measurements can be used to stratify patient populations to ensure that this is achieved. For example, the sex-specific molecular profiles found in some ASD patients indicate that different molecular pathways may be affected in males and females with ASD. This is discussed in further detail in the next chapter which focuses on the effects of sex in psychiatric conditions. However, the finding that ASD males show greater changes in inflammation and ASD females have more changes in hormonal and growth factor pathways suggests that there may be some benefit in targeting these systems in a sex specific manner. In addition, new adjunctive drug treatment strategies could be developed which target co-morbidities such as inflammation or effects of abnormal sex hormone expression using combined treatments with standard medications. For example, some studies have shown that high levels of insulin and androgens are found in women with polycystic ovarian syndrome and treatment with insulin-sensitizing drugs helps to normalize the levels of both hormones and relieves some of the associated symptoms. Thus, biomarker screening of ASD patients, followed by co-administration of drugs which target the affected systems along with existing ASD treatments, could lead to improved patient outcomes.

Chapter 9
Gender and Psychiatric Disorders

The simple fact of whether one is a male or female can be important in the prevalence, incidence, progression and response to treatment for many medical conditions. This includes autoimmune and cardiovascular diseases along with psychiatric disorders. This is perhaps not too surprising since some of the hormonal factors which lead to gender differences during the developmental stages of life are also responsible for the different physiologies in males and females. In turn, these can also confer differential susceptibility or resistance to different diseases. Considering these differences, why are most diseases treated with the same drugs or drug doses independent of gender? This chapter explores such gender differences in healthy individuals with an emphasis on those which lead to differences in physiology, particularly at the level of the brain and synaptic connectivity. This is covered at the levels of anatomy, hormones and other regulatory molecules, and even cultural factors. Next it describes how these factors can contribute to or give rise to different effects and outcomes in diseases such as schizophrenia, depression and autism. Finally, it introduces the concept of potentially developing different treatment approaches in psychiatric disorders based on the gender of the affected individuals. It is anticipated that this will lead to more effective and safer treatments for all subjects affected by these conditions.

The Effects of Gender on Disease

Male and female differences have been widely reported in the molecular processes underlying various medical conditions, ranging from cardiovascular diseases to neurological disorders and psychiatric illnesses. In addition, associated factors such as disease susceptibility, age of onset, progression, death and response to treatment can be influenced by gender. At least some of these effects are likely to be due to sex-specific differences in regulation of the relevant physiological pathways. For example, an elevated activation of the immune system in females has been linked to their

© Springer International Publishing AG 2017
P.C. Guest, *Biomarkers and Mental Illness*,
DOI 10.1007/978-3-319-46088-8_9

increased susceptibility to autoimmune diseases such as multiple sclerosis, rheumatoid arthritis and lupus erythematosus. Sex dimorphisms have also been reported to occur in cardiovascular diseases. For example, myocardial infarctions result in a poorer outcome and higher death rate for females compared to males, although this tends to occur approximately 10 years later in life. At least some of this could be linked to sex differences in the function of heart muscle cells (myocardia) in response to stressful events. In fact, studies in animal models have shown sex-specific differences in the way we respond to stress, which can be seen by activation of different stress-related genes in males and females. The HPA axis can be specifically stimulated in females, most likely due to the known links between the HPA and hypothalamic–pituitary–gonadal axes with the immune system as part of the normal homeostatic maintenance mechanisms (this just means keeping the body fine-tuned).

But is it only females that have it bad? The answer is "no"—not really anyway. Take life expectancy, for example. This has increased with improvements in living conditions over the last century, resulting in a difference between the sexes with females outliving males. In the UK in 2012, the estimated life expectancies for women and men were 82.4 and 78.0 years, respectively. Similar differences exist in other countries. This may reflect a greater susceptibility of men to a higher number of age-related illnesses, such as an earlier appearance of cardiovascular disease, type II diabetes, infection and sarcopenia (loss of muscle mass). The longest unambiguously documented human lifespan occurred for a woman named Jeanne Calment of France (1875–1997), who died at age 122 years and 164 days in 1997 (Fig. 9.1). Madame Calment credited her long lifespan to stress-free living, drinking port wine and consuming large amounts of chocolate. Perhaps paradoxically, she was also an avid smoker for most of her life. The longest undisputed male lifespan was recorded for Jiroemon Kimura of Japan (1897–2013), who died at age 116 years and 54 days in 2013. Taking all of these facts together seems to suggest that females tend to live around 5 % longer than males.

In a similar way, there are likely to be sex-specific responses which could lead to different susceptibilities between males and females in biological pathways which can lead to diseases such as psychiatric disorders. For these reasons, an increased

Jeanne Calment (1875–1997)   Jiroemon Kimura (1897–2013)

Fig. 9.1 The oldest documented female and male in recorded history. On average females live around 5 % longer than males

understanding of sex-differences in the physiological pathways altered in these diseases may also help to identify potential sex-differences which are important in the response to treatments. As an example, administration of growth hormone has a lower efficacy in females compared to males in the treatment of some cardiovascular, immune, metabolic and psychological conditions. This is most likely to result from a relative suppression in growth hormone signalling in females. Therefore, more studies should be performed to elucidate molecular sex differences since this can lead to increased understanding of the biological underpinnings of some diseases and lead to identification of potential new drug targets and development of novel treatment strategies. This is important as there is a current lack using sex-stratified cohorts in clinical trials, which may be a reason for the often spurious and irreproducible findings in such studies.

Are Different Biomarkers Found in Males and Females?

The answer is—yes, of course. The obvious blood based biomarkers are the sex hormones which are markedly different. Most men produce around 12 times more testosterone than women and the levels of oestrogen are higher in women compared to males. Higher levels of testosterone have been linked to physical factors such as brain size and behavioural responses, including aggression and risk taking. It is commonly believed that there is a link between testosterone and aggression although this has not been demonstrated conclusively in scientific studies. The problem of deciphering these studies is that testosterone levels can also be affected by environmental and social influences. Males tend to engage in violent crimes more than females and such behaviour usually begins to emerge in the teenage years around the same time that the testosterone levels rise. Also, other studies have found that testosterone levels are associated with traits such as antisocial behaviour and alcoholism. Of course it is a well known fact that testosterone is an important driver of sexual motivation in males. Males with reduced testosterone often have decreased sexual desire. In contrast, men who have multiple simultaneous sexual relationships tend to have higher testosterone levels compared to men with a single partner or no partner. It has also been shown that oestrogen can affect female sexual desires. As an example, ovulation is associated with increased oestrogen levels and ovulating females tend to display preferences toward more masculine faces and a greater sexual attraction to males who are not their current partner. This does make sense from an evolutionary perspective since the higher oestrogen levels in such females may help them to pinpoint those males with the best genes for procreation.

But there are more molecules that show gender differences apart from testosterone. One study carried out biomarker profiling of serum samples from approximately 200 "normal" males and 200 "normal" females using the multiplexed immunoassay platform in an attempt to identify sex differences in key molecules with specific physiological functions. This led to identification of reproducible sex differences in the levels of 77 out of approximately 250 molecules tested. These included molecules involved in hormone regulation, fatty acid oxidation, immune cell growth and activation, and cell death, which were present at higher levels in females. On the other hand, the molecules that were present at higher levels in males included testosterone or proteins which were mainly involved in immune cell movement. Interestingly, the same study found a deviation in the "normal" sex differences involving molecules related to fatty acid oxidation and hormonal functions in serum samples obtained from Asperger syndrome patients. Females with Asperger syndrome had lower levels of these molecules compared to the typical female controls.

Are There Gender Differences in the Brain?

This is one question that has been asked throughout time: are the male and female brains different? Again the answer is—yes. Without resorting to scientific evidence, these differences may come as no surprise and have been speculated to be along the lines of expected stereotypes (Fig. 9.2).

I won't mention where I stand on these points as I am always a diplomat.

However, the actual differences are subtle. In a landmark study, fMRI analyses performed at the University of Pennsylvania in the USA showed that male and female brains actually have different neuronal connections. In other words, they are wired differently. The fMRI technique is a procedure used to detect regions of brain activity by visualizing and measuring the changes in blood flow (more active regions show higher flow). The University of Pennsylvania study showed that connections within each hemisphere are much stronger in male brains, whereas connectivity between the right and left hemispheres is significantly stronger in female brains (Fig. 9.3). Thus, it could be that the high intra-hemispheric connections seen in males are responsible for the perceived greater abilities of coordinated action. This is because higher connections between the cerebellum and frontal cortex would mean a strengthening of the crosstalk between perception and action, which

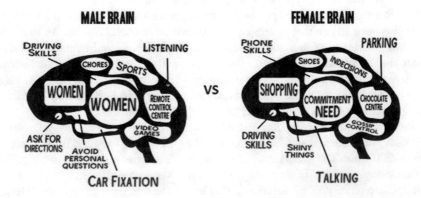

Fig. 9.2 Schematic image showing main functional differences between stereotypical male and female brains

Fig. 9.3 A fMRI study of approximately 1000 brains found that many of the connections in a typical male brain run between the front and back of the same side of the brain, whereas the connections in women run from side to side between the hemispheres

Male
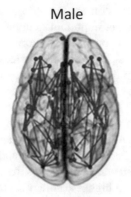

Female
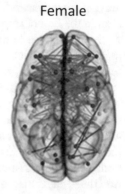

are regulated by those brain regions. On the other hand, the higher inter-hemispheric connections in females are likely to enhance crosstalk of the analytical left side of the brain with the spatial and intuitive functions of the right. As a point of interest, this finding actually supports the old stereotypes suggesting that the male brain is more suited for perception and coordination, and the female brain is better at social skills and multi-tasking. As another point of interest, these male/female differences in neural connectivity were found to be small or non-existent in individuals who were less than age 13 years old with a marked difference occurring after puberty.

Do Brain Structural Differences in Males and Females Relate to Different Behaviours and Characteristics?

Again, the answer is—yes, probably. Studies have also shown that certain areas of the brain have different sizes in males and females. For example, the amygdala is larger in males than in females. This brain region is involved in formation and storage of emotions. Studies in animal models have shown that this larger size is likely to have pronounced effects at the physiological level as male rats have approximately 80 % more excitatory synapses in their amygdalae compared to female rats. This could explain the greater sex drive in males. The sections below indicate the areas where key differences have been reported.

Memory Encoding and Recall

Females have been noted to retain emotional memories more vividly compared to males, which may explain why they are more susceptible to depression (see below). PTSD is also twice as common in females, perhaps for similar reasons. This may be due to hemispheric differences again, this time in the way that emotional memories are encoded and processed. In males, this is carried out on the right side of the amygdala whereas it is performed on the left side in females. The same pattern is seen in the hippocampus, which is important in cognition and memory formation. Activation of this brain region appears to be more dominant on the right side in males and on the left in females. This could be at least one reason why males use fewer verbal approaches in cognitive thinking, compared to females. Another fMRI study was set up to if different regions of the brain are active in males and females when testing working memory. This investigation found that the caudate and lateral prefrontal and parietal cortices were active at similar levels in males and females, and the more difficult tasks led to activation of more areas of brain tissue. However, the left hemisphere had greater activity in females and males showed activation in both hemispheres. Another interesting finding was that females showed slower reaction times but came through with higher accuracy on most tasks.

Intelligence

Gender differences in intelligence have been long debated but results of any studies which have carried out have been largely inconsistent. In general, most studies agree that the mean intelligence levels are approximately the same although males show more variability, with a higher proportion appearing in the top and bottom of the IQ distribution compared to females.

Personality

It is possible that the observed gender differences in brain connectivity and structure could have effects on personality traits. For example, females report that they have higher neuroticism, agreeableness, warmth and openness to feelings whereas males describe themselves as having higher assertiveness and openness to ideas. The same studies showed that males varied more in these traits and those in highly developed world regions reported themselves to be less neurotic, extraverted, conscientious and agreeable than males in less developed countries. In contrast, females self reported more consistency in personality traits in different countries. Finally, the same study suggested that males may have evolved to take more risks and be more socially dominant due to environmental pressures, whereas women evolved to be more cautious and caring. Again, this may be in line with the above mentioned stereotypes.

Occupational Preferences

A number of studies have concluded that males prefer working with things and objects, and females have a preference for working with people. Rankings in specific categories revealed that males had stronger "realistic" and "investigative" interests in engineering, science and mathematics pursuits, whereas women had more interests in "artistic", "social" and "conventional" matters. As stated in the chapter on autism, females tend to perform better than males in tests related to emotions, such as the ability to interpret facial expressions and empathy. A study showed that gender stereotypes are influential when people are asked to judge a hypothetical situation involving conflict. In such scenarios it is usually suggested that females respond with higher levels of emotion and men respond with more anger. However, it is possible that such differences are social constructs and not biologically determined. In other words, boys and girls may learn these responses through education, either directly or indirectly.

Sex Differences in Psychiatric Diseases

Schizophrenia

Although it has been hotly debated, there appears to be no confirmed difference in prevalence of schizophrenia between males and females. However, sex differences have been observed in other aspects of this disease. For example, most researchers agree that there are differences in age of onset as males normally develop the illness earlier between the ages of 18–25 years, whereas the average age of onset is 25–35 years for females (Fig. 9.4). Potential differences in symptoms between males and females with schizophrenia have also been described but the results are not conclusive. Most studies have found that male patients exhibit more negative symptoms and females have a higher prevalence of depressive and anxiety symptoms. In patients admitted with first episode schizophrenia, the majority of females present with more anxiety, illogical thinking and bizarre behaviour compared to males. Considerable evidence based on the study of post-mortem brain tissues suggests that there are deficits in neuronal pathways in schizophrenia. Neuronal impulses travelling from or to the frontal cortex are relayed through the thalamus and several studies have identified structural and functional abnormalities in neuronal fibres interconnecting these regions in schizophrenia patients. For this reason, a study using quantitative PCR measured the expression of 10 genes in five different regions of the thalamus from 14 schizophrenia patients and 16 normal control post-mortem brain samples. This showed that the levels of the myelin sheath associated proteins 2',3'-cyclic nucleotide 3'-phosphodiesterase (CNP) and myelin-associated glycoprotein (MAG) were highly correlated with one another and both were more highly expressed in females compared to males in all thalamic divisions.

A number of molecular profiling studies have now been carried out based on the likelihood that the above sex-related differences may be mirrored by differences in the underlying molecular pathways. Based on the fact that the onset is later in females, most researchers have concluded that males are either more susceptible or that females may be more resistant to neuronal changes. Therefore, the sex hormones have been the primary target of study. For example, oestrogen has been

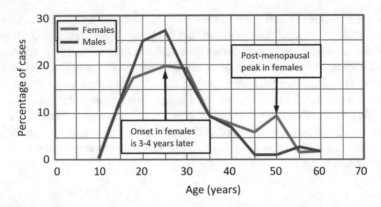

Fig. 9.4 Gender differences in age of onset of schizophrenia may be due to protective effects of oestrogen

shown to have structural, functional and molecular effects in the brain, leading to protective effects in female schizophrenia patients. In line with this, psychosis has been associated with lower oestrogen phases in females and add on estradiol treatment has resulted in improvement of some symptoms. Conversely, testosterone may have deleterious effects on the brain and increase vulnerability to schizophrenia, although the results of investigations which have tested this possibility have not been conclusive. A study using animal models of schizophrenia showed that treatment with estradiol, but not testosterone, led to normalization of the behavioural deficits.

The possibility that sex hormones have a protective role against the pathophysiology of schizophrenia has also been investigated by testing the effects of augmentation treatment with either oestrogens, selective oestrogen receptor modulators (SERMs), testosterone, dehydroepiandrosterone (DHEA), pregnenolone and oxytocin. Significant effects of oestrogens on total, positive and negative symptoms were found in the case of premenopausal women. Likewise, the SERM compound raloxifene was effective in treating the total and negative (not positive) symptoms in postmenopausal women. In contrast, testosterone augmentation was beneficial only for negative symptoms in males and no overall effects were found for DHEA, pregnenolone or oxytocin. These findings suggested that molecules which target the oestrogen pathway could be effective augmentation approaches for treatment of females with schizophrenia. However, considerable further work is required to identify potential side effects of all of these approaches.

Considering these sex-related differences, it is possible that males and females can be affected by different subtypes of schizophrenia or they may exhibit different symptom profiles. Attempts have been made to test these possibilities at the clinical and developmental levels but only one study so far has attempted to characterize the underlying molecular pathways that may underlie these differences. Ramsey and co-workers carried out multiplexed immunoassay profiling analyses in a multi-centre study, comprised of 4 independent cohorts of 133 antipsychotic naive, first episode schizophrenia patients and 133 healthy control subjects and they found something interesting. The concentrations of 16 molecules associated with hormonal, inflammation and growth factor pathways showed significant sex differences in schizophrenia patients compared with controls. In female patients, the inflammation-related analytes alpha-1-antitrypsin, B lymphocyte chemoattractant (BLC) and interleukin-15 had negative associations with symptom scores (PANSS). In male patients, the hormones prolactin and testosterone were negatively associated with symptom ratings. These findings indicated that effects on different physiological pathways may be associated with distinct symptom profiles in males and females with schizophrenia. The same study also investigated a subset of 33 patients before and after 6 weeks of treatment with antipsychotics and found sex-specific changes in the levels of the hormones testosterone and follicle stimulating hormone, and the inflammation-related proteins serum glutamic oxaloacetic transaminase, interleukin-13 and macrophage-derived chemokine. This suggested that males and females may have responded differently to the treatment through changes in different molecular pathways.

Depression

Depression is more common in females than in males (Fig. 9.5). One in four women will require treatment for depression at some time, whereas only one in ten men will need this. The reasons are unclear but may be due to both social and biological factors. A study of 364 males and 751 females with a current diagnosis of major depressive disorder using the NESDA cohort described earlier showed that females had a younger age of onset, higher incidence of panic disorder and life-time occurrence of anxiety disorder, compared to males. At the personal level, one reason that females experience these extra difficulties might be related to the fact that they encode emotional memories more vividly than men as stated above. Also, females tend to be more involved than males in relationships and therefore suffer more when these fall into disorder. Females will also see a doctor more readily if they have symptoms of depression and are thus more likely to receive a depression diagnosis. Also, depression in men may be under-diagnosed because they normally only present to their doctors with other symptoms and are less likely to admit what they are experiencing.

Fig. 9.5 Approximately twice as many females will experience depression in their lifetimes

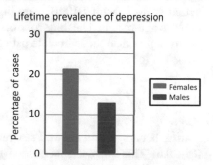

At the molecular level, hormonal differences are usually given as the most likely explanation. This is because females experience greater fluctuation in hormone levels as part of their monthly cycles, as well as during other female-specific life stages such as child birth and menopause. In addition, they are more likely to suffer from hypothyroidism and seasonal affective disorder (SAD). Thus, most investigations into the sex bias of this condition have focused on differences in HPA axis responses to stresses and how the sex hormones can affect this. Of course the cyclic changes in hormone levels in females are thought to be a major factor. During early adulthood, the prevalence of depression increases with a typical onset in the mid 20s. Females of childbearing age are at increased risk and those undergoing menopause are at the highest risk. In support of these observations, studies have shown that aged female rats have the highest susceptibility to develop stress-induced depression and administration of oestrogens produces antidepressant-like effects. Also clinical studies have shown that application of oestrogens can reduce various perimenopausal symptoms. Females also show a greater response to SSRIs, which suggests that antidepressants may have gender-specific responses, most likely due to ovarian hormone effects on the serotonergic system that the SSRIs target. The potential role of testicular hormones has not been considered since these are relatively constant and decrease with age, although aged males respond less well to antidepressant treatments.

Sex differences in incidence of depression arise at adolescence and coincide with androgen and oestrogen levels. Also, oestrogen and androgen signalling in the limbic regions of the brain may influence regulation of HPA axis function. These brain regions appear to be involved in regulation of motivation, emotion, learning and memory. Numerous brain changes occur across the lifespan, including a synaptic reorganization at puberty, which interact to regulate HPA reactivity and therefore modulate behaviour in adulthood. Activational effects of estradiol at puberty may modulate serotonergic circuitry in the amygdala and other structures in a way that places females at greater risk of developing early depression.

Two other molecules have been identified as showing sex differences in depression. The first of these is the inflammatory biomarker C-reactive protein (CRP) which has also been used to predict cardiovascular disease events. One study of 390 participants in a weight loss intervention trial found that symptoms consistent with major depression were significantly associated with CRP levels in males, even after adjusting for differences in age, obesity class, metabolic variables and medications known to affect inflammation. Another study found that the serum levels of the neural growth factor S100B were higher in patients with major depression compared to age-matched healthy controls and the levels in female patients were significantly higher than those in male patients.

Autism

Autism occurs with a skewed sex ratio, as males are affected more than females (Fig. 9.6). This suggests that excessive prenatal testosterone activity could be a risk factor for these conditions, as proposed in the extreme male brain hypothesis (see a

Fig. 9.6 More males than females are affected by autism-related disorders. PDD-NOS = pervasive developmental disorder-not otherwise specified

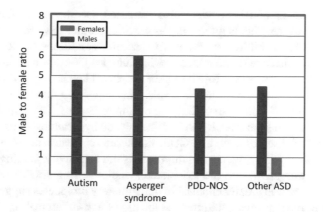

description of this in the previous chapter). In animal models, studies have confirmed that early exposure to high testosterone levels can lead to sex differences in brain structure, cognition and behaviour. Although these effects have been easy to confirm in human males, it has been more difficult in females. It is known that mothers of females with autism have a higher proportion of testosterone-related medical conditions, such as polycystic ovarian syndrome (PCOS). However, the mechanism of how high testosterone exposure in utero can lead to autism may be different in males and females. This has been shown in recent study which assessed the exposure to testosterone in Japanese people with autism using the ratio of the second to fourth digit length (2D:4D) as an estimate of prenatal sex hormone activity. The theory suggests that the higher exposure the fetus has had to testosterone, the more so-called masculine traits the adult will have and the longer the fourth finger will be. Although the study confirmed that males with autism tended to have lower right-hand 2D:4D ratios compared to control males, the right 2D:4D ratios in autism females were higher relative to those of control females. Therefore, the theory may only work in the case of males.

Other molecules which might contribute to sex-related differences in autism have also been identified. For example, a recent study identified the *MAOA* gene as a possible ASD susceptibility factor in males and not in females. *MAOA* encodes the enzyme monoamine oxidase A, which is involved in the synthesis of neurotransmitters such as serotonin. Other studies have profiled serum from adult males and females with Asperger syndrome using multiplexed immunoassays and mass spectrometry. These studies found reproducible sex differences in female Asperger patients in the levels of serum molecules involved in fatty acid metabolism and hormone regulation. In contrast, male Asperger patients had changes in more molecules involved in the inflammation pathway. Specifically, free testosterone levels were elevated in Asperger syndrome females, which provides molecular evidence of the extreme male brain theory. The increased testosterone levels in Asperger syndrome females was paralleled by increased levels of luteinizing hormone in the same subjects, consistent with previous studies which have suggested that perturbed luteinizing hormone pulsatility may predispose or cause hyperandrogenism (most hormones are release in pulses as opposed to a constant release pattern). The serum levels of BDNF were also increased specifically in females with Asperger syndrome. This is interesting as previous studies have found that increased BDNF levels in children with autism spectrum conditions can be associated with atypical brain development (see previous chapter). Serum from females with Asperger syndrome was also found to have decreased levels of growth hormone with a concomitant increase in insulin levels. This suggested that the Asperger syndrome females may also have an underlying insulin resistance. This could be important considering the link between hyperinsulinemia and hyperandrogenism found in previous studies of diseases such as PCOS. Furthermore, treatment of PCOS patients with insulin sensitizing agents can improve insulin sensitivity, leading to alleviation of hyperandrogenism and the associated symptoms. In a similar way, the finding that males showed greater changes in inflammatory molecules suggests that one possible treatment avenue could include the use of anti-inflammatory agents. Taken together, the sex-specific molecular profiles found in Asperger syndrome indicate that different developmental or compensatory mechanisms may occur in males and females with these conditions and therefore different targeted treatments should be considered.

The Future: Should We Consider Sex Differences When Treating Patients with Psychiatric Disorders?

The key point described in this chapter is that sex differences in human behaviour and physiology exist and these can be important in determining the prevalence, severity, subtype and disease course of various psychiatric disorders and for developing more effective therapeutic approaches. In line with this, the National Institute of Health (NIH) in the USA has stated that it will develop policies that address potential sex differences in preclinical research. In terms of behaviour, males and females appear to have different behaviours which are likely to be linked, at least in part, to the different hormonal influences on brain structure and neuronal network development. Males and females also have different responses to stressors which could be a major factor leading to increased susceptibility to disease. For example, females tend to remember emotional traumas more vividly than males which may be one factor that makes them more susceptible to depression. At the molecular level, oestrogen may have protective effects on brain function, leading to a decreased vulnerability of females in the development of conditions such as schizophrenia. In depression, oestrogen treatment may be efficacious as a way of modulating mood during periods of decreased oestrogen levels and increased vulnerability to dysphoria, as occurs in postmenopausal females. In diseases such as autism, males tend to show more significant changes in biomarkers related to inflammation whereas females show alterations in hormonal and growth factor biomarkers. Thus, those female autism patients who have high testosterone levels may be candidates for treatment with anti-diabetic drugs which can lower the levels of this androgen. Considerable further work in this area is required to help overcome the difficulties of treating males and females suffering from a given mental illness under identical conditions. However, the findings in this chapter have stressed the point that the search for novel biomarkers and new drug targets might be facilitated by prior stratification of psychiatric patients into male and female subgroups and this could also lead to the development of better targeted treatment approaches for people suffering from mental disorders.

Part III
Neurodegenerative Disorders

Chapter 10
Biomarkers and New Treatments for Alzheimer's Disease

Alzheimer's disease is a progressive, irreversible neurodegenerative disorder which disrupts thinking, concentration and memory in the affected individuals. Although existing therapeutics can help to improve some of the symptoms there are still no treatments available to halt or delay the incessant progression towards cognitive decline. Currently, more than 46 million people worldwide live with this disease and most of these are over the age of 65 years. The economic impact is huge with a worldwide cost estimated at almost one trillion dollars in the USA alone. Much of this could be averted if early diagnosis and better treatments options were available. Towards this end, this chapter looks at recent developments in diagnostic criteria and identification of peripheral biomarkers under development for early diagnosis of this high burden disease. There is now an attempt at moving beyond the current diagnostic approaches which are based mostly on clinical and neurological examinations. Clinicians today have access to sensitive tests which can reveal cognitive impairment before the disturbances reach the level of dementia and before brain structural damage has already occurred. New plasma and serum biomarkers that have been proposed are based on hallmark pathophysiological features of Alzheimer's disease such as the formation of central amyloid plaques and neurofibrillary tangles which disrupt neurotransmission, along with those associated with increased inflammation, insulin resistance, vascular complications, perturbed oxidative stress and abnormal lipid metabolism. These analytes are currently undergoing testing as biomarker algorithms in combination with clinical and imaging data in attempts to achieve an earlier diagnosis. In this way, disease modifying treatments can be initiated early to delay or prevent progression to the full disease before too much neuronal damage has already occurred. This will be facilitated by co-development of novel treatments which can slow or even prevent disease onset.

© Springer International Publishing AG 2017
P.C. Guest, *Biomarkers and Mental Illness*,
DOI 10.1007/978-3-319-46088-8_10

What Is Alzheimer's Disease?

Alzheimer's disease is the most prevalent form of dementia in the elderly, affecting approximately 0.6 % of the world population and more than 6 % of the population over 65 years old. As people are living longer and longer, this percentage is expected to increase such that it will affect around 1 % of the world population by the year 2030. Alzheimer's disease is characterized by chronic degeneration of cortical neurons, resulting in the loss of memory, cognitive decline and behavioural effects. During the disease course, specific proteins build up in the brain to form plaques and tangles that disrupt connections between neurons and over several years this leads to neuronal death and loss of brain tissue. This is accompanied by a reduced supply of acetylcholine and other neurotransmitters, which are absolute requirements in the functioning of the brain and the whole body. Eventually, this degeneration will lead to death of the individual with a life expectancy of 3–9 years after diagnosis.

In many cases of Alzheimer's disease, the first noticeable symptoms are lapses in memory. In particular, sufferers may have difficulty in recalling recent events. This occurs because one of the regions damaged in Alzheimer's disease is the hippocampus, which plays an important role in short term memory formation (Fig. 10.1).

Interestingly, long term memories do not appear to be affected in the early stages of the disease. Alzheimer's related memory loss can interfere increasingly with daily life as the disease progresses. Below are listed the three stages and associated symptoms (some or all of these may occur):

Early Stage Symptoms (2–5 Years)

1. Difficulties in coming up with the right word or name
2. Repeating one's self in conversations
3. Problems remembering names when being introduced to new people
4. Difficulties in performing tasks in social or work situations

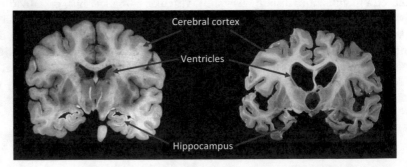

Fig. 10.1 Post-mortem brain cross section showing shrinkage of the hippocampus and effects on other brain regions in Alzheimer's disease

5. Not remembering details of information that has just been read
6. Losing or misplacing objects
7. Increasing problems with planning and organization skills and problem solving
8. Getting lost in a familiar place or on a familiar journey
9. Problems judging distance, navigating stairs or parking the car
10. Forgetting important dates such as birthdays and anniversaries
11. Feeling increasingly moody, depressed or withdrawn
12. Being aware of this decline and possibly attempting to hide it

Middle-Stage Symptoms (2–10 Years)

1. Forgetting events regarding one's past
2. Feeling withdrawn, moody or angry in challenging situations
3. Being unable to recall one's address or telephone number
4. Being confused about one's location, time or date
5. Trouble controlling one's own bladder and/or bowels
6. Undergoing changes in sleep patterns and behaviour
7. Occasionally wandering off and becoming lost
8. Experiencing changes in behaviour including anxiety, compulsivity and depression
9. Becoming increasingly dependent on assistance

Late-Stage Symptoms (1–3 Years)

1. Requiring high levels of full-time care
2. Losing awareness of one's recent experiences and environment
3. Becoming increasingly disorientated
4. Loss of physical abilities, such as walking, sitting and swallowing
5. Having increasing difficulties in communicating
6. Becoming increasingly vulnerable to infections
7. Experiencing delusions and/or hallucinations
8. Experiencing aggression and violent episodes

For most individuals who develop Alzheimer's disease, the onset occurs after the age of 65 years. However, there is an early-onset form which can develop in people who are much younger than this. As with most conditions, the development of Alzheimer's disease has been linked to a combination of factors including advancing age and being female (two-times as many females over the age of 65 develop the disease compared with males). The effect in females may be due changes incurred after menopause and driven by deficits of the hormone oestrogen. In addition, genetics has been implicated as a factor in this disease although the studies on this have been inconclusive and the percentage of inherited cases so low that that this does not seem to be a major component. Finally, the presence of a number of medical conditions such as diabetes, obesity, stroke and cardiovascular disorders are known to increase the chances of developing Alzheimer's disease. In contrast, people who have led a healthy lifestyle in terms of diet, exercise and minimizing alcohol consumption are less likely to develop the disorder.

What Is Going on in the Brain in Alzheimer's Disease?

The pathophysiological hallmarks of Alzheimer's disease are the presence of extracellular amyloid plaques and intracellular neurofibrillary tangles, which can both disrupt neurotransmission (Fig. 10.2). The amyloid protein consists of a peptide called β-amyloid (Aβ) that is generated by proteolytic processing of a larger precursor molecule known as the amyloid precursor protein. Abnormal processing of this precursor can result in different versions of the Aβ peptide, consisting of either 40 or 42 amino acids. The $A\beta_{1-42}$ peptide form is "sticky" and leads to the formation of the toxic insoluble plaques in the brain (Fig. 10.3). This can occur rapidly during the early stages of the disease (Fig. 10.4). The neurofibrillary tangles consist of paired helical strands of a highly phosphorylated micro-tubule-associated protein called tau. The deposition of tau tangles inside neurons most likely leads to an abnormal structure and eventual neuronal death.

There is also a link of Alzheimer's disease to metabolic disorders. Epidemiological studies have found that people with type 2 diabetes mellitus have a high risk of developing Alzheimer's disease. This is because the well known effects on inflammation, oxidation and perturbed metabolism that are seen in diabetes can also lead to cerebrovascular complications and breakdown of the blood–brain barrier. This results in

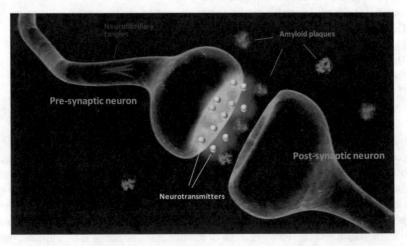

Fig. 10.2 The development of amyloid plaques and neurofibrillary tangles disrupts neurotransmission in Alzheimer's disease

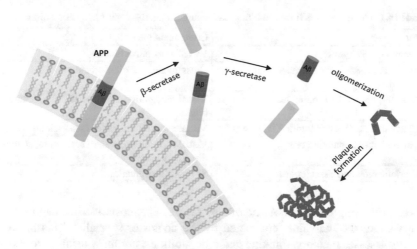

Fig. 10.3 Proteolytic cleavage of the amyloid precursor protein (APP) to produce the "sticky" toxic $A\beta_{1-42}$ peptide, which oligomerizes to form the plaques

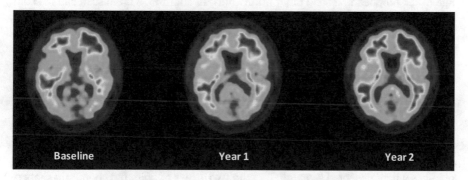

Fig. 10.4 PET scans taken at baseline, 1 year and 2 years showing progressive deposition of amyloid (*red*) in an early-stage Alzheimer's disease patient. The tracer used was Pittsburgh compound B which directly detects amyloid deposits

increased activation of stress-related pathways in the brain which, in turn, promotes some of the key pathological features of Alzheimer's disease, such as mitochondrial energy dysfunction, insulin resistance and generation of amyloid plaques.

How Is Alzheimer's Disease Diagnosed?

As with most neurological disorders (or any disease for that matter), early diagnosis of Alzheimer's disease could have many benefits. For example, it would provide a physiological explanation for the symptoms and it may give early access to treatment, advice and support, all of which could slow the progression of the disease.

Table 10.1 Cognitive tests linked to biomarker changes in early-stage Alzheimer's disease

Test	Component measured
Memory Capacity test	Verbal associative binding
Face Name Associative Memory Exam (FNAME)	Cross-model associative binding
Short-Term Memory Binding test	Visual recognition, change detection, feature binding
Behavioural Pattern Separation-Object test	Visual recognition, pattern separation
Spatial Pattern Separation task	Visual recognition, pattern separation, spatial discrimination
Discrimination and Transfer task	Spatial discrimination

There is no single test for Alzheimer's disease. Diagnosis usually begins by an interview of the individual conducted by the doctor or specialist. This interview usually assesses memory, thinking and other abilities, normally using a written test format. In some cases, the patient may be given a MRI or CT (computerized tomography) brain scan, which may show effects on the substantia nigra region or shrinkage of the hippocampus and other brain areas. In addition, a brain scan can rule out other possible causes such as stroke or tumour development in the brain.

There are a number of tests which measure subtle cognitive changes which have actually been associated with biomarker changes (cerebrospinal fluid amyloid and tau levels, amyloid plaque load, hippocampal shrinkage and certain genetic polymorphisms) found in preclinical Alzheimer's disease (Table 10.1). For example, the Face Name Associative Memory Exam (FNAME) is based on the idea that memory retrieval involves certain networks which may be disrupted in the early Alzheimer's disease. In this test, subjects are shown a series of pictures of unfamiliar faces with names indicated underneath and then asked to recall these. The Short-Term Memory Binding test is a recognition task that examines the subject's ability to identify changes when presented sequentially with two arrays of coloured geometric shapes. In the Spatial Pattern Separation task, the subject is shown an image such as a small triangle and then asked to recall its location on a second image containing two or more similar triangles.

Famous People with Alzheimer's Disease

Listed below are several famous individuals who have suffered or who are still suffering from Alzheimer's disease. Please keep in mind that a multitude of names have not been included on the list for the sake of brevity.

SPORTS
Betty Schwartz (August 23, 1911–May 18, 1999)
American athletics 100 m sprinter, winner of gold medals at the 1928 and 1936
  Olympic games

Ferenc Puskás (April 1, 1927–November 17, 2006)
Hungarian football player and manager, regarded as one of the greatest players of
  all time
Gordie Howe (March 31, 1928–June 10, 2016)
Canadian ice hockey player, widely regarded as the most complete player to play
  the game
Sugar Ray Robinson (Walker Smith Jr) (May 3, 1921–April 12, 1989)
American welter and middle weight boxer, cited as the best pound-for-pound boxer
  of all time

POLITICIANS AND WORLD LEADERS
Harold Wilson (March 11, 1916–May 24, 1995)
British Prime Minister from 1964–1970 and 1974–1976
Margaret Thatcher (the "Iron Lady") (October 13, 1925–April 8, 2013)
British Prime Minister from 1979 to 1990 and first and only female to have held the
  office
Ronald Wilson Reagan (February 6, 1911–June 5, 2004)
American politician and actor, 33rd Governor of California and 40th President of
  USA (1981–1989)

ENTERTAINMENT INDUSTRY
Casey Kasem (April 27, 1932–June 15, 2014)
American disc jockey, radio personality and voice actor, best known for hosting the
  American Top 40
Charles Bronson (Charles Dennis Buchinsky) (November 3, 1921–August 30,
  2003)
American actor best known for his roles in *The Great Escape*, *The Magnificent 7*
  and *The Dirty Dozen*
Charlton Heston (John Charles Carter) (October 4, 1923–April 5, 2008)
American actor best known for his roles in *The 10 Commandments*, *Ben-Hur* and
  Planet of the Apes
Jack Lord (John Joseph Patrick Ryan) (December 30, 1920–January 21, 1998)
American actor best known for his role as Steve McGarrett in *Hawaii Five-O* from
  1968 to 1980
Mike Frankovich (September 29, 1909–January 1, 1992)
American producer of more than 30 films including John Wayne's last movie *The
  Shootist*
Peter Falk (September 16, 1927–June 23, 2011)
American actor known worldwide for his role as Lt. Columbo in the TV series
  Columbo (1968–2003)
Rita Hayworth (Margarita Carmen Cansino) (October 17, 1918–May 14, 1987)
American actress best known for the film *Gilda* and as a top pin-up girl for GIs dur-
  ing World War II
Otto Preminger (December 5, 1905–April 23, 1986)
Austrian–American director best known for his film noir mysteries *Laura* and
  Fallen Angel

MUSICIANS
Aaron Copland (November 14, 1900–December 2, 1990)
American composer best known for his work in the areas of ballet, conducting and film scores
Glen Campbell (born April 22, 1936)
American country music star who appeared solo and with famous groups such as The Beach Boys
Perry Como (Pierino Ronald Como) (May 18, 1912–May 12, 2001)
American singer/entertainer best known as a "crooner" with a career spanning more than 50 years

ARTISTS AND WRITERS
E.B. White (Elwyn Brooks White) (July 11, 1899–October 1, 1985)
American writer best known for his classic books *Stuart Little* and *Charlotte's Web*
Iris Murdoch (July 15, 1919–February 8, 1999)
Irish author famed for several works including the 1978 Booker Prize winning novel *The Sea, the Sea*
Norman Rockwell (February 3, 1894–November 8, 1978)
American painter best known for his works *Rosie the Riveter* and the *Willie Gillis* series
Ross MacDonald (December 13, 1915–July 11, 1983)
American–Canadian author best known for his series of novels featuring the detective Lew Archer
Terry Pratchett (Terence David John Pratchett) (28 April 1948–12 March 2015)
English author best known for his fantasy series of 41 novels about "Discworld"
Willem De Kooning (April 24, 1904–March 19, 1997)
Dutch–American abstract artist best known for his association with the New York School of artists

How Is Alzheimer's Disease Treated?

There is currently no cure for Alzheimer's disease but there are many ways of improving the lives of individuals with this disorder. This includes receiving support, medications, nutritional changes and undertaking activities. The main FDA-approved drugs in use for Alzheimer's disease consist of three main types (Table 10.2). These either work by (1) blocking the breakdown of acetylcholine via inhibition of cholinesterase, (2) stimulating acetylcholine synthesis or (3) blocking glutamate release through inhibition of NMDA type glutamate receptors.

Donepezil, galantamine or rivastigmine are normally used to treat patients in the early of middle stages of Alzheimer's disease. Treatment with these drugs may help patients with memory problems, improve concentration and focus, and help with some aspects of daily living. Memantine may be prescribed for those patients in the middle to late stages of the disease to improve mental abilities and daily living, and to help reduce negative behaviours such as agitation, aggression and delusions. In

Table 10.2 FDA-approved drugs for the treatment of Alzheimer's disease

Drug	Class	Mechanism of action
Donepezil	Cholinesterase inhibitor	Prevents acetylcholine breakdown in the brain
Galantamine	Cholinesterase inhibitor/acetylcholine agonist	Prevents acetylcholine breakdown and stimulates acetylcholine production
Memantine	NMDA receptor antagonist	Blocks toxic effects of excessive glutamate and regulates glutamate synthesis
Rivastigmine	Cholinesterase inhibitor	Prevents acetylcholine and butyrylcholine breakdown in the brain
Tacrine	Cholinesterase inhibitor/ acetylcholine agonist	Prevents acetylcholine breakdown and stimulates acetylcholine production

addition, antidepressants such as fluoxetine may be prescribed for those patients suffering from depressive or anxiety symptoms. Alternatively, counselling may be helpful for some of these patients.

There are also many ways for patients to cope with memory difficulties such as through establishing daily routines or using smartphone apps to set reminders. It is also important that Alzheimer's disease patients keep up with their daily activities as there is plenty of evidence which shows that both mental and physical exercise can lead to some improvements. In fact, clinical studies have found that regular physical activity can minimize the risk of developing Alzheimer's disease and improve cognitive performance in those individuals who are already suffering through the early or middle stages of the disease.

Known Biomarkers for Alzheimer's Disease in Body Fluids

Established biomarkers for diagnosis and monitoring of Alzheimer's disease consist of measuring various proteins and peptides in cerebrospinal fluid via immunological and proteomic approaches as well as by imaging brain features using PET and MRI.

Cerebrospinal Fluid

Since the cerebrospinal fluid is in direct contact with the central nervous system, the molecular content of this body fluid can theoretically provide a useful gauge of brain function. Cerebrospinal fluid can be collected by lumbar puncture of the test subject, which is usually carried out using a local anaesthetic to minimize discomfort. The levels of $A\beta_{1-42}$, tau and hyperphosphorylated tau can be measured by a variety of methods in the cerebrospinal fluid, and can serve as highly specific biomarkers for risk of developing Alzheimer's disease as well as for diagnosis and monitoring disease progression. In general, higher $A\beta_{1-42}$ levels tend to be associated with greater impairments in cognition and studies have found that just measuring this peptide alone can discriminate Alzheimer's disease patients from controls and individuals with mild cognitive impairment, at approximately 85 % accuracy. Other studies have shown that amyloid plaque formation is linked to the presence of apolipoprotein E, which is a major lipid transport protein associated with amyloid deposition and injury repair in the brain. In addition, individuals carrying a particular polymorphism in the apolipoprotein E gene are at higher risk of developing Alzheimer's disease. While cerebrospinal fluid levels of $A\beta_{1-42}$ appear to correlate better with early disease stages before significant symptoms have developed, the levels of tau and hyperphosphorylated tau appear to correlate better with the later stages of the disease which are associated more with synaptic dysfunction. Other proteins and protein fragments have also been identified in cerebrospinal fluid as potential biomarkers of Azheimer's disease. These include the precursor protein chromogranin A and proteins which are involved in stabilizing synaptic structure and neurotransmission, such as the neurofilaments and glial fibrillary acidic protein.

One research group carried out a mass spectrometry-based metabolomic profiling investigation and identified 80 metabolites which were present at different levels in cerebrospinal fluid taken from mild cognitive impairment patients, Alzheimer's disease patients and controls. Of note, more molecular differences were identified in females compared to males. These researchers also found that an increased cysteine–uridine ratio was the best separator of the Alzheimer's disease and control groups at approximately 75 % accuracy. Since uridine is a precursor of membranous signalling protein phosphatidyl choline, the decreased levels of this molecule are thought to be indicative of neurodegeneration. Another mass spectrometry based approach identified a pattern of ten metabolites which could be used to predict progression of Alzheimer's disease with 90 % accuracy. These molecules were arginine, carnitine, choline, creatine, dimethylarginine, histidine, proline, serine, suberylglycine and valine. In addition, the concentrations of the oxidative stress markers 3-nitrotyrosine, 8-hydroxy-2'-deoxyguanosine and isoprostanes have been found to be increased in the cerebrospinal fluid of Alzheimer's disease patients, which could indicate disruption of this pathway. Finally, the finding of increased cerebrospinal fluid levels of norepinephrine in patients may reflect dysfunction of this neurotransmitter in the brain.

Blood, Serum and Plasma

As lumbar puncture is a mildly invasive procedure, biomarkers for Alzheimer's disease have also been carried out in the more accessible media of serum and plasma. One multiplex immunoassay profiling study of serum reported the identification of 18 analytes consisting of haematopoiesis, inflammation and growth factor-related molecules that could be used to distinguish Alzheimer's disease patients from controls with approximately 90 % accuracy. The same panel could also be used to predict conversion of presymptomatic individuals to Alzheimer's disease over a 2–6 year follow-up period. Another multiplex immunoassay study of plasma developed a panel comprised of 30 biomarkers which also appeared useful for distinguishing patients from controls. In these two studies, three proteins with inflammation-related functions were found to be important in the performance of the test. These were angiopoietin 2, pulmonary and activation-regulated chemokine (PARC) and tumour necrosis factor-α. This supports the idea that some of the earliest changes in Alzheimer's disease could be linked with heightened inflammation. In addition, one research group found that two molecules, apolipoprotein E and complement factor H could be used to predict conversion from a prodromal state to confirmed dementia.

A mass spectrometry based metabolomics study found that the plasma levels of glycerophosphocholine and D-glucosaminide could be used to separate Alzheimer's disease patients from controls. The levels of glycerophosphocholine were also elevated in cerebrospinal fluid from Alzheimer's disease patients relative to controls, suggesting that changes in this metabolite may be indicative of pathways disrupted in the brain. Another research group profiled lipids from peripheral blood and found that a set of 10 of these could be used to predict conversion to either mild cognitive impairment of Alzheimer's disease over a 2–3 year period with 90 % accuracy. These changes could reflect the breakdown of cellular membranes involved in an early phase of Alzheimer's disease development.

Micro RNAs (miRNAs) are small non-coding RNA molecules that participate in regulating the production of whole batteries of mRNAs and thereby the proteins that these encode. Many research groups have now reported that specific miRNAs are dysregulated in the blood of Alzheimer's disease patients and many of these molecules have been used as part of panel tests that yield high accuracy for distinguishing patients from individuals with other neurological disorders as well as controls.

Exploring New Treatments for Alzheimer's Disease

There have been many clinical trials attempting to identify new treatments for Alzheimer's disease, although only the acetylcholinesterase inhibitors and the single NMDA receptor antagonist mentioned above have received FDA approval. Nevertheless, several clinical trials are either underway or planned in the testing of several novel therapeutic approaches. All of these pathways have been implicated by the extensive biomarker studies carried out in this area. Most importantly, the hope is that at least some of these new treatment schemes could help to slow or prevent the progression of the disease. Below is a brief description of some of the most promising approaches.

Inhibitors of Amyloid Plaque Production

A large number of compounds have been tested that target the various enzymes (β-secretase and γ-secretase) involved in proteolytic processing of the amyloid precursor protein, which leads to generation of the toxic $A\beta_{1-42}$ peptide (see Fig. 10.3). Many inhibitors of β-secretase have already been approved for use in diabetes research, such as pioglitazone. This drug was shown to reduce $A\beta$ production as well as the insulin resistance that is sometimes found to occur in Alzheimer's disease. A clinical trial also showed that administration of pioglitazone led to increased cerebral blood flow and improved cognitive testing in patients with mild to moderate Alzheimer's disease. Numerous clinical studies have already been undertaken testing the effects of the second enzyme involved in $A\beta$ production, the γ-secretase. Although these drugs showed initial promise, many were withdrawn and trials aborted due to risks of off target side effects. This is mainly due to the fact that γ-secretase is involved in proteolytic activation of multiple important biomolecules. Thus, getting around this problem was deemed too tricky to address. Finally, vaccines which are capable of reducing the $A\beta$ load are currently undergoing clinical testing with some initial promise.

Inhibitors of Tau Tangle Formation

A vaccine called AADvac1 which inhibits formation of tau build up in neurons is currently undergoing testing in early clinical investigations. Thus, far the drug has passed through safety testing in phase 1 clinical trials.

Inhibitors of Cholesterol Synthesis

Since high cholesterol levels have been linked to amyloid deposition, a number of clinical studies have now been undertaken test the effects of cholesterol lowering statins on Alzheimer's disease endpoints. It is anticipated that this class of compounds may also be helpful through their ability to decrease inflammation and increase blood flow. However, the studies carried out thus far have been inconclusive.

Anti-inflammatory and Antioxidant Compounds

Based on the compelling evidence that inflammation and oxidative reduction processes are disturbed in Alzheimer's disease, several drugs have been investigated which target these pathways. Anti-inflammatory agents such as cycloxygenase

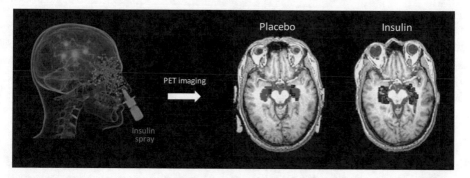

Fig. 10.5 Alzheimer's disease patients were administered either insulin or placebo via an intranasal spray and synaptic activity (*red*) in the hippocampus (*blue*) was determined by positron emission tomography (PET). The image shows greater activity in a patient who received insulin

inhibitors and glucocorticoid steroids have been tested but these approaches showed little efficacy and some adverse effects. Another anti-inflammatory agent which has been approved for arthritis, etanercept, is undergoing testing in clinical trials. In a similar manner, the anti-inflammatory/antioxidant compound curcumin has shown neuroprotective functions such as preventing tau aggregation and stimulating neurogenesis. Therefore, clinical trials investigating the effects of this compound in Alzheimer's disease are also underway.

Drugs Which Target Insulin Resistance

There is now considerable evidence suggesting that insulin signalling pathways support healthy brain function. Given that insulin resistance is a hallmark feature of Alzheimer's disease, a number of anti-diabetic drugs have been tested which enhance insulin action such as pioglitazone, as mentioned above. In line with this, intranasal insulin administration is currently being tested in multiple studies with anticipated positive effects on memory and daily function in mild to moderate Alzheimer's disease patients (Fig. 10.5).

Caffeine

A number of studies have now been carried out which suggest that higher coffee consumption is associated with reduced risk for Alzheimer disease. However, further randomized trials and cohort studies are warranted to confirm this and to determine whether or not there is an association between caffeine intake and the decline in cognition in Alzheimer's disease.

Diet and Exercise

Finally, there is increasing evidence that specific changes in diet towards healthier eating as well as increased physical exercise may be beneficial in the prevention of Alzheimer's disease. There are now two medicinal foods on the market for this purpose, which are available by prescription (CerefolinNAC and Axona). In addition, the traditional Mediterranean diet has shown promise for reducing the risk of developing senile dementia and for decreasing symptoms after diagnosis. However, larger clinical studies are needed to determine whether or not this really works. Other potential non-drug treatments include physical exercise, which is well known for its beneficial effects. It is already known that regular exercise helps to reduce oxidative stress and inflammation, and can also enhance insulin signalling. In addition, exercise is also known to increase neurogenesis, which could be protective in Alzheimer's disease. A clinical trial is now underway to test the effects of aerobic exercise in sedentary elderly patients with mild cognitive impairment on cognition, using cerebrospinal fluid biomarkers and MRI as endpoints.

Future Directions

Alzheimer's disease strikes an increasing number of people in their later years of life and this is expected to escalate as people live longer and longer lives. There are a number of ways the disease is detected but it is typically noticed first as lapses in memory and cognitive functioning. Despite years of investigation, only a few reliable biomarkers have been identified but these may prove to be important since catching the disease early can lead to better treatment options and improved patient outcomes. A number of possibilities are emerging as potential new biomarkers such as the detection of changes in the levels of amyloid peptides and antioxidant-related molecules in cerebrospinal fluid, as well as the monitoring of perturbed patterns of inflammatory molecules and lipid profiles in the peripheral circulation. Also, novel therapeutic options are becoming available such as immunization against the deleterious amyloid plaque formation, treatment of insulin resistance with anti-diabetic drugs, and anti-inflammatory therapies to moderate the heightened state of inflammation. However, undertaking certain preventative measures may be a better option in some cases. There is now considerable evidence that a healthy lifestyle, including a balanced Mediterranean-style diet and physical exercise may delay or even prevent the onset of the disease. This means that if we treat our bodies better through a proper diet and routine exercise, the mind should stay intact and functional for longer periods throughout our lives. Only time will tell.

Chapter 11
Parkinson's Disease, Biomarkers and Beyond

Parkinson's disease is a common neurodegenerative disorder caused by a loss of specific neurons in the brain which ordinarily release the neurotransmitter dopamine. The main symptoms of Parkinson's disease are involuntary shaking or tremor of specific parts of the body, slowed movements and stiff and inflexible muscles. A person suffering from this disease can also experience a range of symptoms associated with other mental disorders such as anxiety, sleep loss, depression and memory loss. At the pathophysiological level, Parkinson's disease is characterized by toxic aggregation of specialized proteins such as α-synuclein in the brain and this occurs in large makeshift organelles known as Lewy bodies. This results in localized inflammation in the brain and all of these signs together can occur several years before the onset of symptoms. Such early changes open up a window of opportunity to identify both biomarkers and novel disease targets for treatment that may moderate or even block disease progression.

What Is Parkinson's Disease?

Parkinson's disease is diagnosed increasingly with age and occurs in approximately 1 % of individuals over 65 years old. It is an incurable and progressive neurodegenerative disorder that slowly leads to a gradual loss of motor control and other functions. Parkinson's disease is a sporadic condition characterized by degeneration of dopaminergic neurons in the nigrostriatal pathway of the brain (Fig. 11.1). This results in underproduction of the vital neurotransmitter dopamine in this region of the brain. The main features of the disease are not obvious until substantial degeneration has occurred and these signs include body tremors, slowed movement, muscle stiffness and rigidity and poor posture. In addition, there are systemic effects with non-motor symptoms often preceding the overt clinical signs, including

Fig. 11.1 Brain sagittal section illustrating the nigrostriatal dopaminergic pathway disrupted in Parkinson's disease

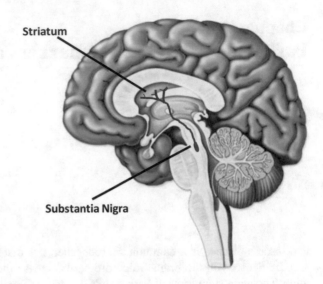

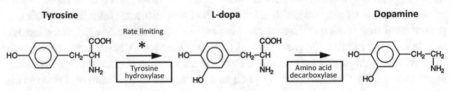

Fig. 11.2 Conversion of the amino acid L-tyrosine to L-dopa and dopamine

dysfunction of the sense and autonomic perturbations such as constipation, along with sleep and cognitive disturbances. Of course the presence of these early symptoms supports the whole body concept of neurological disorders and this also indicates the likely presence of peripheral biomarkers far before onset of the overt disease. Therefore, the high rate of misdiagnosis during these early stages could be improved through the development of suitable biomarker tests. A correct and early identification would thus allow the application of neuroprotective therapeutic approaches long before any irreversible neuronal damage has occurred.

How Is Dopamine Synthesized?

Since the lack of dopamine-producing neurons is the hallmark of Parkinson's disease, it is important to know how this neurotransmitter is normally synthesized in the brain (Fig. 11.2). The rate limiting step in this process is the hydroxylation of the amino acid L-tyrosine by tyrosine hydroxylase to produce an intermediate molecule called L-dopa (catechol di-hydroxyphenylalanine; also known as levodopa). After this, the L-dopa is rapidly decarboxylated to generate dopamine. L-dopa was made famous in the 1990 Penny Marshall film "Awakenings". In this film, a famed neurologist played by the late Robin Williams administers L-dopa to catatonic patients (one of these was played by Robert de Niro) with some positive, albeit temporary, results.

What Causes Parkinson's Disease?

Why Parkinson's disease occurs and how the relevant dopaminergic neurons become impaired, is still not clear. However, one potential explanation is that it is inherited. However, this appears to occur in only a small percentage of cases in which specific genetic abnormalities have been found. There is also evidence that certain toxins that selectively destroy the dopaminergic neurons may trigger the disease. These include manganese, carbon monoxide, carbon disulfide and some pesticides. In addition, there were several cases in the 1980s of people who injected themselves with a synthetic form of heroin containing a compound called MTPT (1-methyl-4-phenyl-1,2,3,6-tetrahydropyridine), which led to an immediate and irreversible form of Parkinson's disease. Another possible cause is that oxidative stress combined with the body's inability to fight this with advancing age may be to blame. In this case, oxidative damage could run unchecked leading to the gradual destructive processes in the brain and other parts of the body. A more controversial link to Parkinson's disease is that it can result from head trauma that can occur through many routes, such as contact sports. There is now a condition called dementia pugilistica that is thought to affect some boxers, wrestlers and other athletes who had suffered repeated blows to the head and/or concussions throughout their careers. Because the condition is thought to affect around 20 % of boxers, there has even been a movement among the medical profession to ban the sport. There is also a related condition chronic traumatic encephalopathy associated with contact injuries incurred by American football players. This featured in the recent film called *Concussion* featuring Will Smith as Dr. Bennet Omalu, a Nigerian forensic pathologist who fights against the National Football League's attempts to bury research on this condition.

How Is Parkinson's Disease Diagnosed?

Diagnosis of Parkinson's disease is usually carried out based on clinical history and physical examination with misdiagnosis occurring frequently when the patient is in the earliest stages of the disease. In clinical practice, it is difficult to distinguish idiopathic Parkinson's from the atypical and vascular forms of the disease, as well as from drug-induced forms, essential tremor and other movement disorders. It is normally classified as early onset Parkinson's disease if the symptoms begin to occur in the individual before the age of 50 years. It is classified as late onset if it develops after this age. The late onset form is the most common and the risk of developing the disease increases with age.

There are a number of clinical rating scales for diagnosis and staging of Parkinson's disease. These include the Unified Parkinson's Disease Rating Scale (UPDRS), the International Classification of Functioning, Disability, and Health (ICF) and the Hoehn and Yahr scale.

Unified Parkinson's Disease Rating Scale

This system is considered as the gold standard for determining Parkinson's disease severity and staging and it is divided into four main areas:

1. Non-motor experiences of daily living (this is a patient self report)
2. Motor experiences of daily living (patient self report)
3. Motor examination (this is scored by the healthcare provider)
4. Motor complications (patient self report)

International Classification of Functioning, Disability and Health

In 2001, the World Health Organization (WHO) endorsed the use of this method as the international standard to describe and measure health and disability. It aims to provide an understanding of the disease consequences from the following perspectives:

1. Body structure and function (rated from zero to complete impairment)

   Nervous system
   Eyes, ears and related structures
   Voice and speech
   Cardiovascular, immune and respiratory systems
   Digestive, metabolic and endocrine systems
   Genitourinary and reproductive systems
   Movement
   Skin and related structures
   Other body structures

2. Activity and participation (rated from zero to complete difficulty)

   Learning and applying knowledge
   General tasks and demands
   Communication
   Mobility
   Self care
   Domestic life
   Interpersonal interactions and relationships
   Major life areas

Community, social, and civic life
Other activities and participations

3. Environmental factors (rated from no barriers to complete barriers)

Products and technology
Natural environment and human-made changes to the environment
Support and relationships
Attitudes
Services, systems, and policies
Other environmental factors

Hoehn and Yahr Staging Scale

This scale is widely used to describe the stages of symptom progression in Parkinson's disease (rated from 1 to 5):

1.0 — Unilateral involvement only
1.5 — Unilateral and axial involvement
2.0 — Bilateral involvement without impairment of balance
2.5 — Mild bilateral disease with recovery on pull test.
3.0 — Mild-moderate bilateral disease and postural instability but physically independent
4.0 — Severe disability but still able to walk and stand unassisted
5.0 — Confinement to bed or wheelchair unless aided

Schwab and England Activities of Daily Living Scale

This system assesses daily activities of the patients in terms of speed and independence using a scale divided into 10 % increments starting at 0 % (bed-ridden and lack of vegetative tasks such as swallowing, bladder and bowel control) to 100 % (complete independence in all activities without slowness, difficulty or impairment).

What Is Going on in the Brain's of People with Parkinson's Disease?

As stated above, Parkinson's disease is characterized by the loss of dopaminergic neurons in the substantia nigra, which leads to the hallmark motor symptoms of bradykinesia, tremor, rigidity and unstable or abnormal posture. Taking advantage of the effects on dopamine (Fig. 11.3), neuroimaging studies using PET, SPECT and MRI have all yielded important information regarding effects on brain structure and function, as well as on disease severity in Parkinson's disease. One approach is

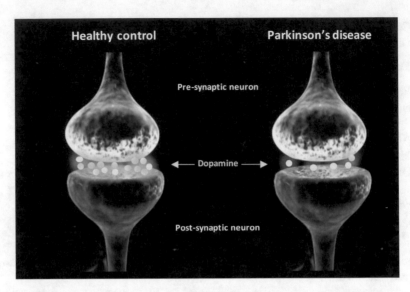

Fig. 11.3 Parkinson's disease is characterized by progressive losses in neurotransmission of dopamine leading to movement disorders and other symptoms

the use of a radiotracer called fluorodopa in PET scans to mimic uptake of L-dopa into the relevant brain regions. Another technique involves scanning for dopamine transporters (DaT) using the DaTSCAN imaging method which uses ^{123}I-ioflupane as a radiotracer in SPECT analyses. Both approaches show an impaired functioning of the striatal region of the brain in Parkinson's disease (Fig. 11.4). In addition, a number of post-mortem studies have been performed in efforts to identify the molecular underpinnings of these characteristic changes. These investigations have zoomed in on the key pathological features of Parkinson's disease, including the degeneration of specific neurons in the substantia nigra. These neurons are marked by the presence of intrusive irregular intracellular organelles known as Lewy bodies (Fig. 11.5). These organelles contain mainly the small regulatory protein ubiquitin and α-synuclein, a protein normally localized to the synaptic buttons where it is involved in regulation of synapse remodelling. Other pathologies which have been identified include changes in mitochondrial function, increased oxidative stress, dysfunction of lysosomes, impaired protein degradation, inflammation and glial cell activation. Of course, further study of these pathways may lead to identification of novel biomarkers as well as the development of appropriate animal models, identification of novel disease targets and the design of new drugs.

Famous People with Parkinson's Disease

The list below includes a number of famous individuals in all walks of life, past and present, who have suffered from or are still suffering with Parkinson's disease. Certainly, there are many others affected by this damaging disease who have not

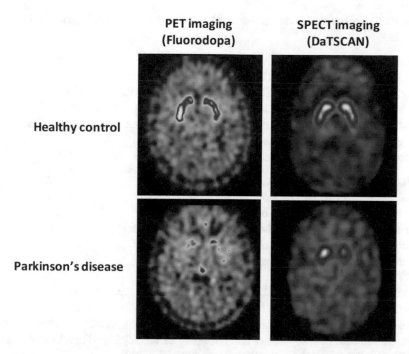

Fig. 11.4 Brain imaging analyses showing impaired function of dopamine pathways in the striatum in Parkinson's disease patients compared with controls. PET analysis using ^{18}F-fluorodopa as the tracer is shown on the right and SPECT analysis using the ^{123}I-ioflupane tracer is shown on the right

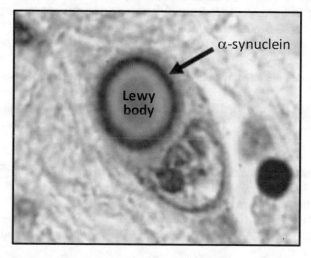

Fig. 11.5 Lewy bodies stained with antibodies against a-synuclein in the brain of patient with Parkinson's disorder

been included but this is due to space limitations that prohibit a more comprehensive list.

SPORTS

John Walker (born January 12, 1952)
New Zealand athlete, first person to run a mile under 3 min and 50 s
Muhammad Ali (Cassius Marcellus Clay) (January 17, 1942–June 3, 2016)
Famous American Olympian and world champion boxer, known as "the greatest"
Roger Bannister (born March 23, 1929)
Neurologist, first person to run a mile under 4 min

POLITICIANS AND WORLD LEADERS

Adolf Hitler (20 April 1889–30 April 1945)
German politician and leader of the Nazi party
Francisco Franco (4 December 1892–20 November 1975)
Spanish general and *caudillo* of Spain
George H.W. Bush (born June 12, 1924)
41st President of the USA
John Paul II (Karol Józef Wojtyła) (18 May 1920–2 April 2005)
Polish Pope and Saint
Mao Zedong (Chairman Mao) (December 26, 1893–September 9, 1976)
Chinese communist party leader and a founder of the People's Republic of China
Pierre Trudeau (October 18, 1919–September 28, 2000)
Canadian politician, 15th Prime Minister of Canada

ENTERTAINMENT INDUSTRY

Billy Connolly (born 24 November 1942)
Scottish comedian and actor
Bob Hoskins (26 October 1942–29 April 2014)
English actor with many staring film roles such as *Who framed Roger Rabbit*
Deborah Kerr (30 September 1921–16 October 2007)
Scottish actress best known for playing the role of Anna in the musical *The King and I*
James Montgomery Doohan (March 3, 1920–July 20, 2005)
Canadian actor best known for playing the character Montgomery Scott "Scotty" in
    Star Trek
Jim Backus (February 25, 1913–July 3, 1989)
American actor best known for providing the voice for the animated character Mr
    Magoo
Michael J. Fox (born June 9, 1961)
Canadian/American actor best known for the *Back to the Future* trilogy
Sir Michael Redgrave (20 March 1908–21 March 1985)
English film and stage actor/director
Vincent Price (May 27, 1911–October 25, 1993)
American actor best known for his laugh in the *Thriller* music video and roles in horror films

MUSICIANS

James Minter (Jimmy) Knepper (November 22, 1927–June 14, 2003)

American jazz trombonist
Johnny Cash (February 26, 1932–September 12, 2003)
American singer-songwriter and guitarist, sold more than 90 million records
Linda Ronstadt (born July 15, 1946)
American popular music singer

ARTISTS AND WRITERS
Mervyn Peake (9 July 1911–17 November 1968)
English writer best known for the *Gormenghast* books
Sir John Betjeman (28 August 1906–19 May 1984)
English Poet Laureate
Steve Alten (born August 21, 1959)
American science fiction author
Salvador Dali (11 May 1904–23 January 1989)
Spanish surrealist painter
Walter Lord (October 8, 1917–May 19, 2002)
American author best known for his non-fiction account of the Titanic sinking, *A Night to Remember*

Perhaps the best known of these was Muhammad Ali who just left us four months ago (at the time of writing) (Fig. 11.6). Ali is still regarded as one of the most significant and inspiring sports figures to have ever lived. In 1999, the BBC (British Broadcasting Company) voted him as the "Sports Personality of the Century". He was also honoured with a star on the "Hollywood Walk of Fame" for his

Fig. 11.6 Potentially the most well-known sports figure of all time, the world champion boxer Muhammad Ali was diagnosed with Parkinson's disease in 1984. At the time, this was suspected as being related to his long career as a pugilist. In 1997, he founded the Muhammad Ali Parkinson Center (MAPC) which combines clinical and research expertise with community support and care

contributions to the entertainment industry in 2002. It is the only star to be mounted vertically in honour of his request that his name should not be walked upon.

Treatment of Parkinson's Disease

To date no modifying or preventative treatments have been identified for Parkinson's disease. This failure is likely to be due to our poor understanding of the multiple pathogenic pathways that can lead to this disorder, which means that existing treatments will not necessarily be targeted in the best way. Also it should be noted that existing and test treatments may ameliorate some of the symptoms without halting the process of neuronal loss. Thus, these are only temporary measures. Since most of the symptoms of Parkinson's disease are caused by a loss of dopamine actions in the brain, many of the currently used drugs are aimed at temporarily replenishing the dopamine supply or mimicking the action of this neurotransmitter in the brain. These types of drugs generally help to improve speed and coordination in patients and they may also reduce muscle rigidity and the characteristic tremor. Peripheral administration of L-dopa is used clinically to increase dopamine concentrations and recent developments have now produced longer acting forms of this compound to make the delivery process smoother and reduce motor fluctuations in the patients. Dopamine concentrations can also be boosted by inhibiting the enzyme that normally breaks it down (this enzyme is called monoamine oxidase) or by enhancing its action through the use of dopamine agonists like apomorphine. Furthermore, psychiatric symptoms such as depression often occur in Parkinson's disease sufferers, most likely through the stress of living with the illness and the mechanical and neurochemical disruption of mood regulating regions of the brain. In these cases, antidepressants many be administered to alleviate some of the symptoms. Although the occurrence of psychosis is rare in untreated Parkinson's disease patients, some of the standard drug treatments which are eventually administered can induce schizophrenia-like side effects such as hallucinations and delusions. In these cases, antipsychotics may be prescribed.

The Need for Biomarkers

There are currently no biomarker tests available that are capable of predicting the onset of Parkinson's disease. This is mostly due to the current dearth of knowledge on the pathophysiological events that occur in this disease which, in turn, makes it difficult to establish diagnostic criteria. Because accurate diagnosis can occur only after a significant degree of neurons have degenerated, there is an urgent need for biomarkers that can identify individuals in a pre-disease prodromal phase or to identify those at most risk of developing the disease. This would enable the application of suitable therapeutics before a significant degree of damage has occurred. The current reliance on clinical assessments to identify Parkinson's disease has hindered research and has led to substantial misdiagnoses when a patient is in the earliest stages of the disease. Furthermore, treatment options are mostly determined using subjective clinical scales that focus on the motor symptoms.

A range of different potential biomarkers for Parkinson's disease have been explored with varying degrees of success. These include information generated by behavioural testing, neuroimaging, genetic susceptibility and sense of smell testing. In addition, some molecular analytes have been identified in serum, plasma and cerebrospinal fluid which may be useful as biomarkers in various aspects of the disease. Of course the best biomarker tests would be cost-effective, non-invasive, user-friendly and fast. The Michael J. Fox Parkinson's Progression Markers Initiative (PPMI) was launched in 2010 and is comprised of a group of 20 institutions in the USA and Europe, with the mission of identifying high quality biomarkers for meeting the objectives outline above. The centres are focussing on enrolling patients in the earliest stages of the disease before any medication has been received. They are looking at movement, cognitive and brain biomarkers, in addition to blood, urine, DNA, and spinal fluid sampling in 400 newly diagnosed PD patients over a 3- to 5-year period. In addition, they are employing standard operating procedures for clinical assessments, sampling and storage, technical analyses and data processing to ensure maximum impact of the findings.

Biomarker Candidates Identified for Parkinson's Disease

Screening for biomarkers employs techniques attempt to identify patterns of variation in behaviour, brain structural changes, genetic markers, mRNA transcripts, proteins and small molecules in biological samples such as living brains, post-mortem tissues, saliva, blood, urine and cerebrospinal fluid.

Behavioural Biomarkers

Non-motor-related symptoms can occur in early Parkinson's disease before the extensive loss of dopamine neurons in the substantia nigra and these changes appear to reflect neural degeneration in other brain areas. This includes disturbances in the sense of smell, sleep behaviour, visuospatial abilities, cognition, mood and behaviour. Therefore, functional tests targeting these symptoms may be used to indicate risk of developing Parkinson's disease years before the overt motor symptoms develop. Assays which have been developed along these lines include the University of Pennsylvania smell identification test (UPSIT), the REM Sleep Behaviour Disorder Screening Questionnaire (SBDSQ), the Bradykinesia Akinesia Incoordination (BRAIN) test and various accelerometry-based gait analyses. These tests have been used in a number of studies and all have shown potential if incorporated into Parkinson's disease screening programmes. Of course it should be noted that the effectiveness of these analyses would be increased by developing risk algorithms comprised of multiple different kinds of tests, including those involving behavioural, imaging and molecular biomarkers.

Transcranial Sonography

Another method to identify individuals with Parkinson's disease is transcranial sonographic imaging of the substantia nigra. This is simply an ultrasound test of the brain and has been drawing a lot of interest recently, considering that it is both easily accessible and cost effective. Several studies using this approach have found that a high proportion of Parkinson's disease patients have an enlargement of a region of the substantia nigra that is thought to be associated with increased iron concentrations there. However, this effect can also be detected in other neurodegenerative diseases and may even be related to ageing. Therefore, further work is required to help pin down the specificity of this test for Parkinson's disease, such as incorporating other measures into a multi-parameter algorithm, as mentioned above.

Genetic Biomarkers

Parkinson's disease is thought to result from interaction of a variety of factors such as genetics, lifestyle and environmental influences. However, genetic investigations have identified only weak associations between single nucleotide polymorphisms in a number of genes and Parkinson's disease development. In addition, genome wide association studies have identified less than 20 genetic factors that may be involved and these cannot adequately explain the heritability of the disease. Furthermore, genetic testing can only identify a trait or susceptibility for Parkinson's disease but these cannot be used for diagnosis or classification of disease stage. However, familial forms can occur that can be explained by mutations in the gene *SNCA*, which encodes the α-synuclein protein. We already know that this protein is important in the disease process and therefore it has been the subject of most investigations attempting to identify biomarkers for early diagnosis.

Serum, Plasma and Cerebrospinal Fluid Biomarkers

The identification of biomarker panels for neurodegenerative disorders like Parkinson's disease has been obstructed by the impracticalities and inconceivability of sampling brain tissues from living individuals, as well as the difficulties in obtaining high quality post-mortem samples from patients after death. The fact that the cerebrospinal fluid comes into direct contact with the brain probably makes this biofluid the most informative for brain abnormalities occurring in Parkinson's disease. However, collection of biosamples from this source is more invasive than taking samples from the bloodstream in the form of serum and plasma. For this reason, these body fluids have become the media of choice for biomarker studies in Parkinson's disease. This also fits with the mind body connection discussed throughout this book.

α-Synuclein

Although many candidates related to specific facets of Parkinson's disease pathophysiology have been tested for their applicability as circulating biomarkers, α-synuclein has emerged as the front runner. Normally, this protein appears to be involved in promoting synaptic vesicle release by interacting with proteins and lipids in the synaptic buttons (see Chap. 1). In the disease state, this protein appears in misfolded and aggregated forms in the Lewy bodies, as described earlier. This transformation may be due to abnormally high phosphorylation of the α-synuclein protein within the neurons and both phosphorylated and non-phosphorylated forms of the protein can also appear in the cerebrospinal fluid as well as in the blood. Amazingly, one study showed that the phosphorylated form of α-synuclein can be diagnostic in blood samples taken at the first presentation of Parkinson's disease patients. Combined with its genetic linkages as described above, this protein is a prime candidate as a biomarker for prediction of Parkinson's disease.

As mentioned above, one school of thought is that Parkinson's disease may actually be detectable in the periphery years before the actual disease is manifested at the clinical level. In fact, it has been suggested that α-synuclein aggregation may begin in other parts of the body before it occurs in the brain. For example, one study found that 9 out of 10 individuals in the early stages of Parkinson's disease had α-synuclein inclusions in the gastrointestinal submucosa. Again, this bodes well that it may be possible to detect biomarkers for prediction of Parkinson's disease at a time when it really matters and potential disease modifying treatments can be attempted.

Inflammation-Related Proteins

A recent study measured the levels of 10 cytokines and C-reactive protein at baseline, 18 months and 36 months in Parkinson's disease patients and age-matched controls. This analysis resulted in identification of tumour necrosis factor-α and the interleukins 1-β, 2 and 10 as being significantly higher in patients. In addition, some of these analytes were associated with more rapid motor progression over 36 months and others were linked with lower cognition scores at all times. However, it should be noted that these findings are still preliminary and require validation studies.

Novel Drug Treatment Approaches in Parkinson's Disease

There are still no treatments for Parkinson's disease that can reduce the persistent neurodegeneration or halt the disease process. However, the recent increase in our knowledge of the disease gained through biomarker-related studies has resulted in the identification of potential novel disease models and therapeutic targets for use in preclinical development. This section discusses the failures and most promising approaches to arise over the last 3 years in clinical investigations.

Failed Approaches

Although the idea of testing neuronal growth factors in Parkinson's disease met with initial excitement, clinical trials involving these compounds have thus far shown little or no improvement in disease progression. In addition, the results of trials investigating the effects of peroxisome proliferator-activated receptor-gamma (PPAR-γ) agonists, coenzyme Q and creatine have been less than satisfactory despite the fact that these compounds are targeted at boosting the failing mitochondrial energy supply. Finally, testing of dopamine receptor agonists such as pramipexole also failed but this time it was due to adverse drug effects.

Methods that Result in Symptom Improvement

Although L-dopa is the most effective medication to improve motor control in Parkinson's disease, the usual route of administration is oral, which requires absorption via the gastrointestinal system. As a result, an inhaled form of L-dopa known as CVT-301 has been tested in clinical trials and found to decrease patient "off" times by 1.6 h. However, further efficacy and safety studies are still underway.

Novel Approaches to Slow or Halt Disease Progression

There are several new approaches which may have disease modifying properties. These are listed below:

1. α-synuclein immunotherapy

    As α-synuclein has been implicated as a key contributor to neuronal death in Parkinson's disease, immune-based therapies targeting this molecule have been tested. Thus far, clinical trials have shown that administration of anti- α-synuclein monoclonal antibodies leads to a reduction in circulating α-synuclein levels, with no notable adverse events. Further trials are now on the books to determine the efficacy of this approach.

2. Caffeine

    Coffee drinking has been associated with a reduced risk of Parkinson's disease and therefore trials have begun testing the efficacy of caffeine in reducing the telltale symptoms such as the motor deficits. Thus far, the trials have shown some success.

3. Inosine

    Epidemiological studies have shown that higher circulating levels of uric acid are associated with reduced risk of developing Parkinson's disease and with slower disease progression. Given that inosine is a uric acid precursor, inosine is now undergoing early testing in clinical studies to determine whether or not it will have disease modifying properties. Given that high uric acid levels can lead to gout, patients who already have high levels of this molecule or a history of gout will be excluded from the trials.

4. Nicotine '

Epidemiological studies have also found that tobacco smoking results in a decreased risk of developing Parkinson's disease. Thus, studies testing the effects of nicotine in patients and controls are underway. Of course these studies will have to consider the potential harmful effects of the tobacco smoke in these studies and potentially use alternative means of delivery.

Other molecules are also undergoing testing as potential treatments for Parkinson's disease, including antioxidant compounds such as glutathione and agents that enhance insulin action such as agonists for the glucagon-like peptide 1 receptor. In addition, screening of drugs that have already been proven to be safe in humans for other indications could identify novel compounds for treatment of Parkinson's disease, given that a suitable preclinical assay is developed for testing. This could include screening for drugs which alter α-synuclein phosphorylation using a cytomics-based approach, as described in Chap. 6.

Future Directions

This chapter describes recent advances using biomarkers for improved diagnosis and classification of individuals with the neurodegenerative disorder Parkinson's disease, which strikes a high proportion of individuals in the later stages of life. At present, the results have been disappointing as only a few biomarkers have been offered up as potential candidates. However, the inconsistent finding that this disease may be detected several years before it has begun to cause damage in the brain has offered a potential lifeline. Even more striking is the likelihood that these biomarkers may be detectable in other parts of the body in addition to the brain. As with the other neurological conditions presented in this book, this concept suggests that a more integrative view of Parkinson's disease may be required, which considers involvement of the entire body, and not just the brain. The translation of biomarkers as user friendly tests of the blood would be ideal to enable early identification of those individuals who have the highest risk of developing the disease. This should be combined in algorithms accounting for other factors such as patient and family history, clinical scores and imaging data to increase the power of the prediction. In this way, patients can be placed on disease modifying therapeutics at a time before irreversible damage to the brain has already occurred. Finally, new treatment strategies may be emerging such as longer acting forms of the traditionally administered drug L-dopa, to increase dopamine action in the brain, and immunotherapies which are aimed at decreasing the levels of the aggregated form of α-synuclein throughout the body. Thus, considering Parkinson's disease as a disorder of the entire body may offer solutions for new biomarkers and better treatment options.

Part IV
The Future

Chapter 12
The Future: Towards Personalized Medicine

In the case of psychiatric disorders, repeated failures in converting scientific discoveries into newer and better drugs have precipitated a crisis and eroded confidence in the pharmaceutical industry. This final chapter describes how future innovations driven by application of biomarkers can encourage the pharmaceutical companies to look beyond the delivery of all-serving blockbuster drugs to more individualized approaches. This will also have influence on the field of psychiatry and result in the delivery of sensitive and specific biochemical tests to complement and increase the power of the traditional questionnaires for improved diagnosis. Furthermore, the application of emerging biosensor tools will facilitate point of care testing by the fusion of biochemical and clinical information. In this way, patient data will be composed of their past medical histories, bio-patterns and prognoses information, resulting in personalized profiles or molecular fingerprints for each individual with these conditions. On the near horizon, the application of mobile communications technology and grid computing will assist disease prediction, diagnosis, prognosis and compliance monitoring. It is expected that such personalized medicine approaches will help to move psychiatric medicine into the twenty-first century and help us to manage these diseases better than we ever have before.

Why Do We Need Improved Technologies in the Study of Psychiatric Illnesses?

Psychiatric disorders are debilitating behavioural and mental health diseases which can strike almost any individual at any age. They can seriously impair quality of life, social well-being and productivity, with significant knock on effects on society and the economy. According to the World health Organization (WHO), mental disorders will be the second leading cause of disability worldwide by the year 2020. Major

© Springer International Publishing AG 2017
P.C. Guest, *Biomarkers and Mental Illness*,
DOI 10.1007/978-3-319-46088-8_12

depression alone affects approximately 20 % of the people in Europe and the socio-economic costs of this disorder approximate 1 % of the gross domestic product, with major costs incurred outside the healthcare systems due to negative impacts on employment issues. According to the Global Burden of Disease study, mental disorders like depression account for over 12 % of years lived with disability (YLD—a measure of the years lost due to living with a disability). This is compounded by the fact that current medications are around only 50 % efficacious and they can also have serious side effects. Despite the urgent medical need, no clinically validated molecular biomarker test exists in a user friendly and rapid format for any psychiatric disorder. Perhaps counterintuitively, pharmaceutical companies have recently cut research in the area of mental disorders, due to outcome drawbacks associated with the lack of understanding of these conditions and the poor availability of surrogate markers to support drug discovery efforts. Obviously, this is a vicious circle that reduces the capacity of these companies to identify new medicines. Thus, there is an urgent need for new technologies and biomarker-based approaches that are fast, cost-effective and simple to use for all operators. The two biggest markets which should be targeted are point of care use by the primary care providers and clinicians to aid diagnosis, and pharmaceutical companies conducting large scale clinical trials and/or testing new drugs. At present, the diagnosis and treatment of major depression is often performed in primary care by general practitioners but with a low success rate since 70 % of these patients are under-treated. As noted in previous chapters, applying the best available therapeutic approach for each patient is paramount. In line with this, patient stratification using biomarkers is essential for correct treatment. Use of accurate in vitro diagnostic lab-on-a-chip devices that can accurately classify patients according to the type of disorder, or disease subtype, will help to reduce duration of illness and improve compliance by placing the right patients on the right treatments as early as possible. This will help to change the overall paradigm from reactive medical care to a more efficient and patient-friendly personalized medicine scheme (Fig. 12.1).

Severe mental disorders such as schizophrenia, major depression and autism affect people of all ages and can be ameliorated only partially, if at all, with prescription medicines. In conventional medical practice, the physician comes to a diagnosis based on observation of symptoms and prescribes a specific treatment, accordingly. However, psychiatrists, mental health workers and physicians can be confronted by a vast array of ethereal and even subtle cognitive, behavioural and emotional disturbances which they aim to classify as a particular syndrome. They can also use other observations relating to the patient, such as appearance and behaviour, self-reported symptoms, mental health history and lifestyle. The most systematic classifications are usually performed via questionnaires based on standard diagnostic criteria, guided by the ICD-10 produced by the WHO and the DSM produced by the American Psychiatric Association (APA), as described earlier. However, in reality, these diagnostic evaluations are often incomplete or they may be carried out in an unstructured way, leaving them open-ended and prone to inaccuracy. Taken together, this can add to the chances of misdiagnosing the patient.

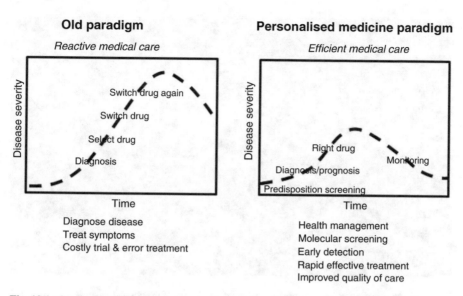

Fig. 12.1 Comparison of the old and new paradigm of medical care as distinguished by the use of biomarkers for improved patient care

One reason for the above difficulties is that the current diagnostic protocols for mental disorders do not normally take into account observations from biological or pathophysiological sources. However, as described throughout this book, recent advances are now beginning to unravel the neurobiochemical nature of these disorders and we are increasing our understanding of them. These advances have included the results of studies through minimally invasive molecular imaging techniques, such as PET, MRI, fMRI, ultrasound and single photon emission computerized tomography (SPECT), and molecular screening approaches such as mass spectrometry and multiplex immunoassay profiling. However, despite being of interest scientifically, such techniques have limited clinical utility due to the resulting low to mediocre sensitivity and specificity in diagnosing or classifying the patients, which is likely to be related to the sheer complexity of these disorders. Thus, molecular biomarkers are essential since the fundamental lesions underlying diseases of the central nervous system are likely to be associated with complex perturbations of physiology, manifested as dysregulation of molecular expression profiles. Consequently, any understanding of these disorders should be accompanied with a systems level view of the dysregulated factors associated with the pathogenesis and should include construction of disease- or symptom-specific molecular biomarker panels. Systems biology is simply a biology-based approach that focuses on complex interactions within biological systems, using a holistic rather than the traditional reductionist approach (looking at the whole rather than the parts). The discovery of authenticated biomarkers could significantly enhance future mental healthcare if the resulting tests can be controlled using standard operating systems and clinical decision making, and the resulting platforms can be established as fast,

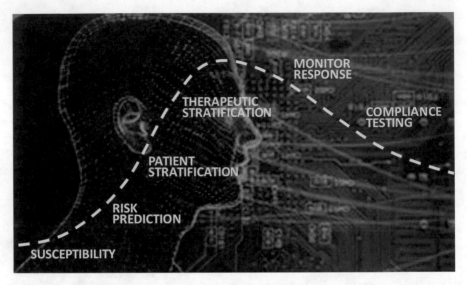

Fig. 12.2 Application of biomarkers in the disease evolution of psychiatric disorders

user friendly devices, which can be used in a point of care setting. Molecular biomarker tests developed in this way should allow precise prediction of many parameters such as disease susceptibility and risk, diagnosis and prognosis, patient and therapeutic stratification, patient response, adverse drug reactions, as well as being useful for monitoring treatment compliance (Fig. 12.2). In addition, using this approach for earlier detection and treatment of a mental illness should help to minimize the adverse physical and emotional effects traditionally associated with long periods of under treatment.

What Kind of Biomarker Tests Should Be Developed?

Although genomic studies are able to identify genes conferring susceptibility to a particular disease, the functional abnormalities are ultimately reflected in the post-translational proteome. This is because the proteome is dynamic whereas the genome is static (apart from mutational changes due to causes such as ionizing radiation and cancer). Recent years have seen the increasing use of proteomics as a tool for the discovery of biomarkers for diagnosis, monitoring disease progression, treatment response and for the identification of novel therapeutic targets. It is also important to remember that analysis of CNS disorders is difficult as the brain is not readily accessible for invasive diagnostic purposes. Thus, sources such as cerebrospinal fluid, serum, plasma, saliva, urine and peripheral blood cells have been targeted as these have a higher accessibility and utility. Serum presents the most accessible fluid for proteome analysis although it is inherently complex, with a dynamic range exceeding ten orders of magnitude and the 22 most abundant proteins accounting for 99 % of the total proteome mass. The concentration of proteins in plasma ranges from as low as 5 pg/mL (interleukin-6—although it is likely that there are other proteins present at even lower undetectable levels) to the very high 50 mg/mL (albumin). Thus, proteomic platforms must cope with the high complexity, wide dynamic concentration range of the targeted proteins and the vast abundance of post-translational modifications of these same proteins. Although the human genome contains around 20,000 genes that encode proteins, the actual number of proteins may be much higher (around 2,000,000) due to post-translational changes of the proteins such as proteolysis, phosphorylation, glycosylation, amidation, and others. Key techniques which have been used to analyze at least some of these proteins have already been described in earlier chapters and application of these platforms in serum and plasma analyses has identified multiple biomarker candidates. It should also be kept in mind that there is no universal platform which can capture the entire proteome and so combinations of techniques will often give a more complete picture in screening studies. However, for ease of use in the clinical environment and to enable point of care testing, a single method is optimal and this should be simple to use by non-experts, rapid and sensitive (see below).

Clinical Impact of Biomarkers

In the near future, it is conceivable that the increasing use of biomarkers by clinicians will lead to more extensive "bio-" signatures in individuals that reflect the molecular changes occurring in health or disease as well as those (good or bad) resulting from medication effects. A carefully selected multiplex diagnostic test should enhance the sensitivity of identifying all patients afflicted with an illness (lower false negatives) and the specificity of correctly identifying those patients who do not have the disease (fewer false positives). Ideally, biomarker tests should have sensitivities and specificities which are both greater than 90 % and these tests should be highly reproducible to be of any real clinical utility. Currently, tests for schizophrenia and depression have almost reached these specifications in limited clinical studies, although they still require multisite validation and repeat studies to be considered for further development. Also, the platforms involved are typically composed of multiple large components, requiring considerable technical expertise in order to operate them and they do return results rapidly enough to be of use in a point of care setting. In fact, most of these platforms would require more than 1 day from the time of a single sample input to the results output.

Given the high prevalence and burden of mental disorders such as major depressive disorder, large markets have emerged for novel and more effective diagnostic and treatment approaches. In line with this, multiplex antibody-based biomarker tests have now been developed on devices which are approximately the size of a credit card. A novel innovative approach towards development of a lab-on-a-chip system for point of care diagnostics was described in 2012 and has been rapidly

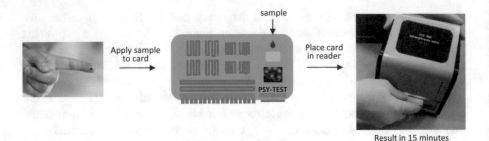

Fig. 12.3 The "lab on a chip" work flow

developing since this time. This platform offers up the possibility of an inexpensive, on-site and user friendly analysis. The system also features a high degree of modularity which allows adaptation of various assay types and formats. Thus far, immunoassay-based microcards have been developed for detection of protein markers such as prostate specific antigen using either electrochemical or optical readouts. This format has several advantages over existing systems, with the main ones being that it is user friendly with no expertise required, and it is rapid with a low cost. A typical test involves application of blood samples to the card, which is then inserted into a small table top book-sized reader to produce diagnosis as a final "score" in less than 15 min (Fig. 12.3). Such devices obviously fit the bill as point of care innovations since they can be used directly in a doctor's office, with the result obtained within that short time frame. Thus, one of the anticipated major benefits will be a rapid turnover time, which will help to cut down on waiting periods for the results of lab tests which can often take several weeks using standard methods. This will help the physician to make informed choices to ensure that the right patients are placed on the right treatments as early as possible. It does not just stop here since these devices can contain a universal serial bus (USB) port for connection to a computer and transmission of data to a smartphone or a wearable device through incorporation of near-field communication (NFC) modules.

Other Benefits: Cost Savings for the Healthcare Systems

Investments in development of user friendly biomarker tests as described above are of high importance considering current costs associated with the medium diagnostic success rate and the less than adequate response to psychiatric medications, as stated in earlier chapters. This is critical since the medical consequences and associated stigma of mental illnesses can create a vicious cycle of discrimination, social isolation, unemployment, drug abuse, long-term institutionalization or homelessness. A user friendly device for early diagnosis would help to alleviate such problems by reducing misdiagnosis, allowing early intervention, and enabling healthcare professionals to decide on the best treatment course. In turn, this would reduce the duration of untreated illness and help to ameliorate disease severity. This supports the WHO Mental Health Action Plan for Europe to promote mental well-being, tackling discrimination, preventing mental health problems, and ensuring access to appropriate primary care. Applying the best available therapeutic approach for each patient is paramount. In line with this, patient stratification using biomarkers is essential to enable the best treatment. The same approach could also be highly useful for pharmaceutical companies conducting large scale clinical trials and testing new drugs. Improved stratification of patients would help to enhance trial outcomes by ensuring that the right patients are enrolled in these studies from the start. Also, specifically designed trials could provide data on treatment response biomarkers and aid in elucidation of novel therapeutic targets, which would lead to development of novel drug concepts.

The deployment of point of care diagnostic tests will help to meet the needs of providing effective therapies at an earlier stage in the disease. It is expected that economic benefits will be seen at individual and societal levels through positive outcomes such as improved health, increased productivity and reduced healthcare costs. The cost of depression alone in the European Union was approximately 43 billion Euros in 2010 and 17 billion (40%) of this was attributed to indirect costs. Assuming that availability of a point of care test for a particular mental illness could increase the number of successfully diagnosed and treated patients by twofold, national healthcare savings of approximately 8.5 billion Euros per year would be realized for the indirect costs alone. The actual savings are likely to be even higher considering that many other disorders can also present with depressive symptoms, which have not been previously recognized by the physician. For example, over 95% of major depression patients present at their clinics with cognitive symptoms which can negatively impact quality of life and ability to function professionally and socially and longer periods of untreated disease can lead to increased debilitation.

Is There a Market for Point of Care Devices?

The gold standard for diagnosis of psychiatric disorders is a structured interview using the appropriate clinical tools, as described in earlier chapters. Such tools are used in clinical and research scenarios to diagnose and stratify patients but are seldom available in primary care units. The avalanche of candidate biomarkers has now created a market for high-throughput multiplexed immunoassays that allow simultaneous quantification of the analytes. Centralized hospital laboratories have evolved into automated systems of barcoded patient identification, sample collection, processing and passage via high-throughput multiplexed clinical chemistry and immunoassay platforms, although sometimes critical results do not emerge or become available until after several days or even weeks later. Such delays can obviously impede a timely diagnosis and decision making by the medical care providers and thereby negatively affect patient outcomes. However, recent years have seen a slow revolution in diagnostic practices, with a trend towards point of care testing. This has been made possible by technological advances in assay performance, instrumentation and miniaturization biosensors and transducers, as well as incorporation of better data processing techniques.

Point of care testing employs a spectrum of devices ranging from large table-top instruments to implanted, wearable and handheld modules. Most of the current handheld systems consist of a disposable reagent pre-filled strip incorporated into a cartridge or a cassette to facilitate addition of the sample and to house the conduction of the test, and the resulting signal is interpreted visually or measured using an inexpensive reader or meter. Point of care tests that are already in existence can be used to measure blood gases and ions, biomarkers of heart attack, cholesterol, markers of diabetes, pregnancy, certain infectious diseases, and the presence and/or levels of alcohol and some drugs of abuse. Such devices usually consist of a test strip or pad, lateral flow devices, cards, slides, flow-through tubes and cassettes with built-in meters and displays. Perhaps the most familiar example is the Clearblue Digital Pregnancy test® marketed by SPD Swiss Precision Diagnostics GmbH, which provides a pregnant/not pregnant readout and an indication of the number of weeks since conception.

The IVD market revenue is expected to reach 75 billion US dollars by the year 2020. The USA is currently the largest market in the world at 24 billion dollars each year. According to the 2014 US Market for the In Vitro Diagnostics Tests industry report, point of care devices also make up the largest part of this at 3.6 billion dollars per year. Growth in this area reflects the preferences of the patient's for being seen doctor's offices or clinics rather than in central laboratories. Since mental disorders are becoming an increasing problem and are expected to lead to the highest burden on the healthcare systems over the next few years, more innovative lab-on-a-chip devices and connectivity solutions will be required that use only small amounts of sample (such as a single drop of blood), with less effort and time spent in performing the test and, of course, all of this has to happen at a lower cost. In addition, large information technology companies including Apple and Google have shown a growing interest in the medical diagnostic market. This is due to the simplicity of connecting an app result with a diagnostic answer using smart software. Likewise, pharmaceutical companies now consider that low cost and time-effective biomarkers are essential for facilitating decision-making in clinical trials and for driving existing and new drug discovery efforts.

How the Evolution of Biosensors Has Aided Development of Point of Care Devices

Biosensors are analytical devices which combine biorecognition of an analyte such as a protein with a suitable physiochemical transducer to convert this into an electrical signal, which in turn is processed to display the results in a user friendly manner (Fig. 12.4). Transducers can exploit any physical signal such as those from

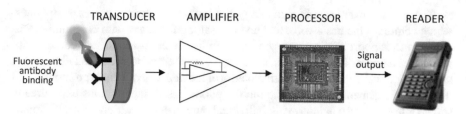

Fig. 12.4 Schematic of a biosensor based on capture of an antigen (*gold*) using an immobilized antibody (*black*) and detection using a tagged secondary antibody (*green*), the physiochemical transducer for conversion of the recognition into an amplified and processed electrical signal, and the instrumentation for outputting the signal in a user friendly format

electrochemical, optical, acoustic, magnetic or thermal sources. The device shown in the schematic shows a transducer linked to the binding of a fluorescent antibody as in the typical sandwich immunoassay format (see Chap. 3 for more details).

Recent trends in biosensor technology include the use of several novel and exciting recognition and transducer systems such as those linked to peptides, imprinted molecules, electro-polymerized and immobilized biomolecules, and nanowires, as well as signal amplification systems that use futuristic magnetic nanoparticles and quantum dots. All of this sounds far off in the future but is here already. It is just a matter of combining these technologies in the best possible way. Other enhancements have come from the development of microfluidics and lab-on-a-chip platforms. Many classes of analytes such as proteins have now been measured employing such platforms, including those using miniaturized on-chip electrophoresis, optical trapping, immunoassay and multiplex immunoassay techniques.

e-Psychiatry

An important requirement of developing point of care devices revolves around the use of mobile communication and the internet so that data can be organized, interpreted, evaluated and formatted for simple presentation to the end users. This would allow testing using real-time, multiplexed sensors, combined with artificial intelligence and mobile communication systems for the analysis and display. Such an approach would be of high relevance in the field of mental disorders, since these tend to be long term debilitating diseases that afflict people worldwide and require constant monitoring and treatment. There is also a movement towards the use of networked computing to aid disease prediction, diagnosis, prognosis and even monitoring of medication compliance. The envisaged result includes the creation of bioprofiles or molecular fingerprints at the individual level that can fuse genomics, proteomics and imaging data with patient histories.

In the twenty-first century, mobile phones have become nearly ubiquitous. At the end of 2011, there were a staggering six billion mobile phone subscriptions, equating to around 87 % of the world population (although many individuals have more than one). A rapid convergence in technology has led to the emergence of smartphones, which combines the basic voice and text messaging functions with computing technologies to support applications, sensing, wireless internet access and connectivity with other devices. These features, combined with their widespread and almost universal usage make them an attractive platform for delivery of health promotion and disease management. In fact, basic mobile phone-based interventions have already shown initial promise in improving individual outcomes in a variety of health conditions, diseases and habit control. A recent review of trials involving health care interventions delivered by cell phones showed improvements in 61 % of the targeted outcomes. These included better attendance of patients at their medical appointments, speedier diagnosis and treatment, and improved communication, along with behavioural improvements such as cessation of smoking and increased medication compliance. There were also clinical enhancements including improved blood sugar control, decreased symptoms of asthma and lower stress levels. Most of these interventions involved automated text or voicemail reminders of things such as appointment dates and times, medication dosing or symptom assessments reminders, or basic encouragement of preventive health behaviours or self-management activities.

Finally, recent trends have even seen the development of multiplex immunoassay based tests on a handheld and cost-effective smartphone-based colorimetric microplate reader platform, which uses 3D-printed opto-mechanical attachments to illuminate small wells embedded in a 96-well plate via a light-emitting-diode optical fibre array. The captured images are transmitted to centralized servers for processing and generation of diagnostic results, which can be delivered to the user within 1 min. Thus far, this mobile platform has been tested in a clinical microbiology laboratory using mumps, measles and herpes simplex I and II virus immunoglobulin tests, and these tests yielded correct identification of the virus with accuracies of over 98 % in all cases. It is not hard to imagine that similar tests for other diseases such as psychiatric disorders will be available in the near future.

Personalized Medicine?

The main theme of this book has been leading to the point that biomarker signatures identified in patient blood samples can be used to improve their lives by enabling personalized medicine approaches. As the name implies, personalized medicine is the situation in which persons are treated based on their individual biomarker profiles, as opposed to being treated as one of many with a particular disease using a one size fits all standard drug. Most recently, such programs have been initiated in the field of psychiatry and neurobiology as treatment of the frequent co-morbidities in patients. In each case, the co-morbidities can be identified by biomarker profiling as a means of stratifying individuals based on those who are likely to benefit most from the treatment. The finding that some schizophrenia patients show more prominent changes in hormones and others show greater alterations in immune-related molecules (described in Chap. 4) suggests that it may be possible to distinguish subgroups of patients based on the molecular profiles and then treat them accordingly to ease symptoms. However, it should be stressed that it is still very early days in these kinds of trials and much further work is needed.

Targeting Insulin Resistance

A number of studies have now established that many first onset schizophrenia patients have high insulin levels and/or suffer from increased insulin resistance and many patients show similar effects after treatment with second generation antipsychotics such as olanzapine. So bearing this in mind, the personalized medicine approach suggests that schizophrenia patients who have high levels of insulin might benefit from add-on treatment with drugs that improve insulin action. Insulin receptors exist on virtually all cells of the body, including the neurons. This possibility has already been explored in Chap. 4. This described the use of the insulin-sensitizing agents to correct the antipsychotic-induced insulin resistance without compromising the psychiatric benefits. This finding paves the way for future studies using biomarker testing for perturbed insulin signalling to stratify patients or to monitor treatment responses or side effects (Fig. 12.5). Also, adrenal steroid hormones have

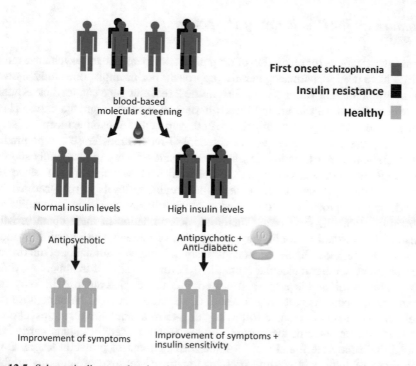

Fig. 12.5 Schematic diagram showing the use of point of care biomarker testing to stratify schizophrenia patients prior to treatment according to the presence or absence of insulin resistance

been tested as a novel means of treating the negative symptoms of schizophrenia as an add-on treatment in medicated patients. Thus, patients with high negative symptom scores may benefit specifically from such an approach.

Targeting Inflammation

Similar to the above approach, it has also been possible to use anti-inflammatory drugs to target patients showing perturbation in their immune or inflammatory profiles. Again, Chap. 4 described the use of anti-inflammatory drugs as a treatment for schizophrenia symptoms. This approach also led to an improvement in negative symptoms. These studies support the case that initial screening of patients for biomarkers related to immune status, followed by an anti-inflammatory in combination with antipsychotic treatment, could lead to improved outcomes. A recent meta-analysis of schizophrenia clinical studies showed that co-administration of non-steroidal anti-inflammatory drugs, such as ibuprofen, naproxen and aspirin, with standard antipsychotics is effective for alleviation of psychiatric symptoms as well as the co-morbid inflammation-related conditions.

Biomarkers for Prediction of Response

Some investigations for prediction of drug prediction response for psychiatric medications have used a pharmacogenomics approach. For example, one study showed that a combination of variants in the histamine 2 receptor gene can predict response to clozapine treatment in around three quarters of the schizophrenia cases. There have also been findings indicating that genetic variants of dopamine receptors, serotonin receptors and enzymes such as catechol-O-methyltransferase can be predictive of clinical response and/or development of side effects. Another study showed that simple measurements of physical parameters such as waist circumference and body mass index, as well as blood tests for circulating levels of triglycerides and high density lipoproteins could be used to predict the antipsychotic-induced development of metabolic syndrome. Therefore, determination of these characteristics ahead of time would be useful for stratification of patients and selection of those who would be most likely to benefit from adjunctive antidiabetic treatments. Measurement of other molecular biomarkers before the start of treatment may also prove to be useful for this purpose. A proteomic study carried out in 2012 found that schizophrenia patients with higher levels of serum prolactin at baseline are more likely to have a better outcome following long-term treatment with antipsychotics. In terms of schizophrenia symptoms, another serum proteomics study carried out around the same time found that baseline measurement of prolactin levels, along with those of fatty acid binding protein, ferritin, C-reactive protein, myoglobin, complement factor H and interleukin-16 could predict improvement in positive symptoms of schizophrenia in response to antipsychotic treatment. Furthermore, measurement of baseline insulin and matrix metalloproteinase 2 levels could be used for prediction of improved negative symptoms (described in more detail in Chap. 4). The same study also suggested that low levels of the hormones insulin and leptin and high levels of the inflammatory growth factor TGF-beta might be useful in the prediction of whether or not subjects would suffer from relapse with a mean time of approximately 16 weeks. However, this latter study group was only composed of 18 patients. Given the potential aid to the attendant physicians and the medical benefits to patient lives, further studies along these lines should be carried out using larger patient groups. Identification of biomarkers resulting from these analyses, if proven accurate, could be used for patient stratification and prediction of response and would therefore be useful for clinicians to make biomarker-guided decisions and take appropriate actions such as making patient assessments based on empirical evidence, counselling and readjustment of medications as necessary.

The Future

The findings in this book describe recent advances using biomarkers which prove the existence and continuous activation of a mind-body connection, which can be used for improved diagnosis and classification of individuals with psychiatric disorders. The ultimate goal is to provide more informed treatment strategies and, therefore, better patient care. The use of multiplex biomarker approaches also provides a means of deciphering the vast array of molecular pathways affected in specific psychiatric conditions as well as in different subtypes of these diseases. Eventually, this could lead to a more holistic view of the perturbed biological pathways in psychiatric diseases, which have now been established to affect multiple organ systems of the body and not just the brain. This can also be seen by the identification of distinctive molecular profiles for mental disorders in body fluids such as cerebrospinal fluid, serum and plasma, as well as cells such as peripheral blood cells and skin fibroblasts, and tissues such as post-mortem pituitary glands. Many patients show distinct patterns of circulating molecules which are indicative of a dysfunctional immune or inflammatory system with elevated levels of inflammatory cytokines. Others have metabolic abnormalities such as altered HPA axis function, high insulin levels or insulin resistance. The finding that schizophrenia patients can have distinct changes in one or more of these pathways suggests that this mental disorder may be composed of different subtypes, at least as seen at the level of circulating biomarkers. This is likely to also be true for other psychiatric conditions such as major depressive disorder, bipolar disorder and autism-related conditions. On the other hand, the distinct biomarker patterns may simply be different manifestations of the same underlying disorder. Remember that it is still not certain if these molecular changes are causative factors or merely a consequence of the disease (the chicken-egg scenario). Nevertheless, better classification of patients based on molecular profiles would enable stratification prior to treatment and this could be particularly useful for guiding physicians as to the type of add-on therapy required to achieve the best possible patient outcomes. Furthermore, the use of multiplex biomarker tests on credit card-sized or handheld devices which are capable of distinguishing schizophrenia subtypes may be useful for rapid identification of patients who are most likely to respond to specific psychiatric medications either alone or in combination with other drugs. It obvious that such an approach could result in more effective treatment of patients with fewer side effects and, thus, a decreased rate of those patients deciding to discontinue medication due to unbearable side effects. For instance, schizophrenia patients with high insulin levels could be administered insulin sensitizing agents as an adjunct therapy with standard antipsychotics, with the aim to minimize metabolic side effects such as weight gain while still achieving improved symptomatic responses.

Stratification of individuals with psychiatric disorders using proteomic biomarker profiles to assign the right treatments to the right patients is the exact definition of the personalized medicine. This approach has already seen increased use in the field of cancer treatment. For example, the measurement of human epidermal growth factor receptor 2 over-expression in breast cancer can be used to identify those women who are most likely to benefit from treatment with Trastuzumab (Herceptin), an antibody-based therapeutic developed by Genentech in the 1990s. Similar efforts in other disease areas such as psychiatric disorders would support the personalized medicine movement by helping to de-convolute the complexity of these diseases, which have heretofore been based solely on symptom characterisation.

Concluding Remarks

This book has described the challenges, recent successes and the anticipated future linked with the production of blood-based biomarker tests for psychiatric disorders. All of this is based on the idea that psychiatric diseases are not simply disorders of the brain but are most likely to be caused by disturbances in whole body networks. This can be seen as two-way communication between the brain and periphery and much of this is mediated through the bloodstream. Thus, the current diagnostic process and strategies for developing novel medicines are finally undergoing a paradigm change after around 100 years of stasis. In addition, the regulatory health authorities, such as the Food and Drug Administration in the USA, now consider that the incorporation of blood-based biomarker tests into clinical platforms and patient trials to be of critical importance for the future of drug discovery and development. In response to this need, they have now called for efforts to modernize the methods, tools and techniques as necessary steps. In addition, there has been a shift towards the development of tests on rapid bench-top or handheld devices so that these can be used in point of care scenarios. However, there are still many challenges ahead. Most of these relate to the fact that we have only just skimmed the surface of the ocean of biomarkers that are likely to be useful for diagnostic purposes. This is mainly due to the fact that most biomarkers have a complex nature

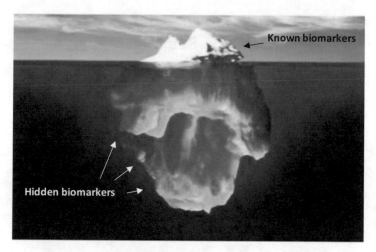

Fig. 12.6 Iceberg analogy of the known biomarkers

and are exceedingly low in abundance, and therefore are still not technologically accessible. One analogy would be that of the proverbial iceberg in which only a small fraction is visible and the remainder is hidden beneath the surface (Fig. 12.6) Delving further into these "deeper" biomarkers will most likely require the development of more reproducible and sensitive methods, as well as the appropriate user friendly platforms to facilitate point of care testing.

Of course, this will require development of new technologies and a massive integration of these as well as the existing ones—a seemingly impossible task. However, there have been other times in history where seemingly impossible tasks have been undertaken and achieved against all odds. One of the best examples is probably the project which put men on the moon in July 1969. The project began in 1962 when President John F. Kennedy said in his address to the nation: "we will go to the moon because it is hard." The interesting thing about this was that in the early 1960s, many of the technologies needed had not even been invented yet. Not only that, a massive integration of technologies was required to produce what many have claimed to be the most important achievement of the twentieth century. Therefore, there is reason to be optimistic that further technological advancements will be made and interdisciplinary approaches will come together to help overcome the current limitations in the field of biomarkers. This will help to guide medicine into the twenty-first century lead to the better clinical outcomes in all major fields, including psychiatry.

Bibliography

Chapter 1

"The blood of the Insane" in the History of Psychiatry [17(4): 395–418 Copyright © 2006. Sage, Thousand Oaks, CA.

Middleton FA, Strick PL. Cerebellar projections to the prefrontal cortex of the primate. J Neurosci. 2001;21:700–12.

Laborit H. On the mechanism of activation of the hypothalamo--pituitary--adrenal reaction to changes in the environment (the 'alarm reaction'). Resuscitation. 1976;5:19–30.

Mastorakos G, Zapanti E. The hypothalamic-pituitary-adrenal axis in the neuroendocrine regulation of food intake and obesity: the role of corticotropin releasing hormone. Nutr Neurosci. 2004;7:271–80.

Straub RH, Buttgereit F, Cutolo M. Alterations of the hypothalamic-pituitaryadrenal axis in systemic immune diseases - a role for misguided energy regulation. Clin Exp Rheumatol. 2011;29:S23–31.

Reagan LP. Insulin signaling effects on memory and mood. Curr Opin Pharmacol. 2007;7:633–7.

Solas M, Aisa B, Mugueta MC, Del Río J, Tordera RM, Ramírez MJ. Interactions between age, stress and insulin on cognition: implications for Alzheimer's disease. Neuropsychopharmacology. 2010;35:1664–73.

Schwarz E, Guest PC, Rahmoune H, Harris LW, Wang L, Leweke FM, et al. Identification of a biological signature for schizophrenia in serum. Mol Psychiatry. 2012;17:494–502.

Chapter 2

Group BDW, Atkinson AJ, Colburn WA, DeGruttola VG, DeMets DL, Downing GJ, Hoth DF, et al. Biomarkers and surrogate endpoints: preferred definitions and conceptual framework. Clin Pharmacol Ther. 2001;69:89–95.

Yung AR, Nelson B, Stanford C, Simmons MB, Cosgrave EM, Killackey E, et al. Validation of "prodromal" criteria to detect individuals at ultra high risk of psychosis: 2 year follow-up. Schizophr Res. 2008;105:10–7.

© Springer International Publishing AG 2017
P.C. Guest, *Biomarkers and Mental Illness*,
DOI 10.1007/978-3-319-46088-8

Cannon TD, Cadenhead K, Cornblatt B, Woods SW, Addington J, Walker E, et al. Prediction of psychosis in youth at high clinical risk: a multisite longitudinal study in North America. Arch Gen Psychiatry. 2008;65:28–37.

American Psychiatric Association. American Psychiatric Publishing. DSM-5. 1000 Wilson Boulevard, Arlington, VA, 2013. pp. 22209–3901.

Van Os J, Hanssenm M, Bijl R, Straus RA. Revisited; a psychosis continuum in the general population? Schiz Res. 1969;2000(29):11–20.

Laughren TP. What's next after 50 years of psychiatric drug development: an FDA perspective. J Clin Psychiatry. 2010;71:1196–204.

Schwarz E, van Beveren NJ, Ramsey J, Leweke FM, Rothermundt M, Bogerts B, et al. Identification of subgroups of schizophrenia patients with changes in either immune or growth factor and hormonal pathways. Schizophr Bull. 2014;40:787–95.

Fillman SG, Sinclair D, Fung SJ, Webster MJ, Shannon WC. Markers of inflammation and stress distinguish subsets of individuals with schizophrenia and bipolar disorder. Transl Psychiatry. 2014;4, e365.

Ryan MC, Collins P, Thakore JH. Impaired fasting glucose tolerance in first-episode, drug-naive patients with schizophrenia. Am J Psychiatry. 2003;160:284–9.

Guest PC, Wang L, Harris LW, Burling K, Levin Y, Ernst A, et al. Increased levels of circulating insulin-related peptides in first-onset, antipsychotic naïve schizophrenia patients. Mol Psychiatry. 2010;15:118–9.

Kim YK, Myint AM, Verkerk R, Scharpe S, Steinbusch H, Leonard B. Cytokine changes and tryptophan metabolites in medication-naive and medication-free schizophrenic patients. Neuropsychobiology. 2009;59:123–9.

Holt RI, Peveler RC, Byrne CD. Schizophrenia, the metabolic syndrome and diabetes. Diabet Med. 2004;21:515–23.

Chapter 3

Ovens J. Funding for accelerating drug development initiative critical. Nat Rev Drug Discov. 2006;5:271.

Johnston-Wilson NL, Sims CD, Hofmann JP, Anderson L, Shore AD, Torrey EF, et al. Disease-specific alterations in frontal cortex brain proteins in schizophrenia, bipolar disorder, and major depressive disorder. The Stanley Neuropathology Consortium. Mol Psychiatry. 2000;5:142–9.

O'Farrell PH. High resolution two-dimensional electrophoresis of proteins. J Biol Chem. 1975;250:4007–21.

Wilkins MR, Sanchez JC, Gooley AA, Appel RD, Humphery-Smith I, Hochstrasser DF, et al. Progress with proteome projects: why all proteins expressed by a genome should be identified and how to do it. Biotechnol Genet Eng Rev. 1996;13:19–50.

Link AJ, Eng J, Schieltz DM, Carmack E, Mize GJ, Morris DR, et al. Direct analysis of protein complexes using mass spectrometry. Nat Biotechnol. 1999;17:676–82.

Lander ES, Linton LM, Birren B, Nusbaum C, Zody MC, Baldwin J, et al. Initial sequencing and analysis of the human genome. Nature. 2001;409:860–921.

Venter JC, Adams MD, Myers EW, Li PW, Mural RJ, Sutton GG, et al. The sequence of the human genome. Science. 2001;291:1304–51.

Chapter 4

American Psychiatric Association. Diagnostic and statistical manual of mental disorders, 4th ed., text revision. American Psychiatric Association, Washington, DC, 2000.

Muller N, Schwarz M. Schizophrenia as an inflammation-mediated dysbalance of glutamatergic neurotransmission. Neurotox Res. 2006;10:131–48.

Hasnain M, Fredrickson SK, Vieweg WV, Pandurangi AK. Metabolic syndrome associated with schizophrenia and atypical antipsychotics. Curr Diab Rep. 2010;10:209–16.

Pennington K, Beasley CL, Dicker P, Fagan A, English J, Pariante CM, et al. Prominent synaptic and metabolic abnormalities revealed by proteomic analysis of the dorsolateral prefrontal cortex in schizophrenia and bipolar disorder. Mol Psychiatry. 2008;13:1102–17.

Guest PC, Schwarz E, Krishnamurthy D, Harris LW, Leweke FM, Rothermundt M, et al. Altered levels of circulating insulin and other neuroendocrine hormones associated with the onset of schizophrenia. Psychoneuroendocrinology. 2011;36:1092–6.

Offord S, Lin J, Mirski D, Wong B. Impact of early nonadherence to oral antipsychotics on clinical and economic outcomes among patients with schizophrenia. Adv Ther. 2013;30:286–97.

Antonova E, Sharma T, Morris R, Kumari V. The relationship between brain structure and neurocognition in schizophrenia: a selective review. Schizophr Res. 2004;70:117–45.

Erritzoe D, Talbot P, Frankle WG, Abi-Dargham A. Positron emission tomography and single photon emission CT molecular imaging in schizophrenia. Neuroimaging Clin N Am. 2003;13:817–32.

Reynolds GP. The neurochemistry of schizophrenia. Psychiatry. 2005;4:21–5.

Zhu YK, Li CB, Jin J, Wang JJ, Lachmann B, Sariyska R, Montag C. The 2D:4D ratio of the hand and schizotypal personality traits in schizophrenia patients and healthy control persons. Asian J Psychiatr. 2014;9:67–72.

Moore L, Kyaw M, Vercammen A, Lenroot R, Kulkarni J, Curtis J, et al. Serum testosterone levels are related to cognitive function in men with schizophrenia. Psychoneuroendocrinology. 2013;38:1717–28.

Zhang XY, Zhou DF, Cao LY, Wu GY, Shen YC. Cortisol and cytokines in chronic and treatment-resistant patients with schizophrenia: association with psychopathology and response to antipsychotics. Neuropsychopharmacology. 2005;30:1532–8.

Taguchi A, Wartschow LM, White MF. Brain IRS2 signaling coordinates life span and nutrient homeostasis. Science. 2007;317:369–72.

Craft S. Insulin resistance and Alzheimer's disease pathogenesis: potential mechanisms and implications for treatment. Curr Alzheimer Res. 2007;4:147–52.

Ben-Jonathan N, Hugo ER, Brandebourg TD, LaPensee CR. Focus on prolactin as a metabolic hormone. Trends Endocrinol Metab. 2006;17:110–6.

Müller N, Riedel M, Schwarz MJ. Psychotropic effects of COX-2 inhibitors--a possible new approach for the treatment of psychiatric disorders. Pharmacopsychiatry. 2004;37:266–9.

Bumb JM, Enning F, Leweke FM. Drug repurposing and emerging adjunctive treatments for schizophrenia. Expert Opin Pharmacother. 2015;16:1049–67.

Schwarz E, Guest PC, Steiner J, Bogerts B, Bahn S. Identification of blood-based molecular signatures for prediction of response and relapse in schizophrenia patients. Transl Psychiatry. 2012;2, e82.

Schwarz E, Steiner J, Guest PC, Bogerts B, Bahn S. Investigation of molecular serum profiles associated with predisposition to antipsychotic-induced weight gain. World J Biol Psychiatry. 2015;16:22–30.

Gebhardt S, Haberhausen M, Heinzel-Gutenbrunner M, Gebhardt N, Remschmidt H, Krieg JC, et al. Antipsychotic-induced body weight gain: predictors and a systematic categorization of the long-term weight course. J Psychiatr Res. 2009;43:620–6.

Perkins DO, Jeffries CD, Addington J, Bearden CE, Cadenhead KS, Cannon TD, et al. Towards a psychosis risk blood diagnostic for persons experiencing high-risk symptoms: preliminary results from the NAPLS project. Schizophr Bull. 2014;41:419–28.

Chan MK, Krebs MO, Cox D, Guest PC, Yolken RH, Rahmoune H, et al. Development of a blood-based molecular biomarker test for identification of schizophrenia before disease onset. Transl Psychiatry. 2015;5:601.

Chapter 5

Patel A. Review: the role of inflammation in depression. Psychiatr Danub. 2013;25 Suppl 2:S216–23.

Lee TS, Quek SY, Krishnan KR. Molecular imaging for depressive disorders. AJNR Am J Neuroradiol. 2014;35(6 Suppl):S44–54.

Ustun TB, Korff MV. Primary mental health services. In: Ustun TB, Sartorius N, editors. Mental illness in general health care: an international study. Chichester: Wiley; 1995. p. 347–60.

Alonso J, Codony M, Kovess V, Angermeyer MC, Katz SJ, Haro JM, et al. Population level of unmet need for mental healthcare in Europe. Br J Psychiatry. 2007;190:299–306.

Leuchter AF, Cook IA, Hunter AM, Korb AS. A new paradigm for the prediction of antidepressant treatment response. Dialogues Clin Neurosci. 2009;11:435–46.

Papakostas GI, Fava M. Does the probability of receiving placebo influence clinical trial outcome? A meta-regression of double-blind, randomized clinical trials in MDD. Eur Neuropsychopharmacol. 2009;19:34–40.

Duman RS, Malberg J, Nakagawa S, D'Sa C. Neuronal plasticity and survival in mood disorders. Biol Psychiatry. 2000;48:732–9.

Guest PC, Knowles MR, Molon-Noblot S, Salim K, Smith D, Murray F, et al. Mechanisms of action of the antidepressants fluoxetine and the substance P antagonist L-000760735 are associated with altered neurofilaments and synaptic remodeling. Brain Res. 2004;1002:1–10.

Duman RS. Synaptic plasticity and mood disorders. Mol Psychiatry. 2002;7 Suppl 1:S29–34.

Jokinen J, Nordstrom P. HPA axis hyperactivity and attempted suicide in young adult mood disorder inpatients. J Affect Disord. 2009;116:117–20.

Rasgon NL, Kenna HA. Insulin resistance in depressive disorders and Alzheimer's disease: revisiting the missing link hypothesis. Neurobiol Aging. 2005;26 Suppl 1:103–7.

Häfner S, Baumert J, Emeny RT, Lacruz ME, Thorand B, Herder C, et al. Sleep disturbances and depressed mood: a harmful combination associated with increased leptin levels in women with normal weight. Biol Psychol. 2012;89:163–9.

Olusi SO, Fido AA. Serum lipid concentrations in patients with major depressive disorder. Biol Psychiatry. 1996;40:1128–31.

Maes M. A review on the acute phase response in major depression. Rev Neurosci. 1993;4:407–16.

Dorn LD, Dahl RE, Birmaher B, Williamson DE, Kaufman J, Frisch L, et al. Baseline thyroid hormones in depressed and non-depressed pre- and early-pubertal boys and girls. J Psychiatr Res. 1997;31:555–67.

Karege F, Vaudan G, Schwald M, Perroud N, La Harpe R. Neurotrophin levels in postmortem brains of suicide victims and the effects of antemortem diagnosis and psychotropic drugs. Brain Res Mol Brain Res. 2005;136:29–37.

Maes M, Meltzer HY, Cosyns P, Suy E, Schotte C. An evaluation of basal hypothalamic-pituitary-thyroid axis function in depression: results of a large-scaled and controlled study. Psychoneuroendocrinology. 1993;18:607–20.

Stefanescu C, Ciobica A. The relevance of oxidative stress status in first episode and recurrent depression. J Affect Disord. 2012;143:34–8.

Papakostas GI, Shelton RC, Kinrys G, Henry ME, Bakow BR, Lipkin SH, et al. Assessment of a multi-assay, serum-based biological diagnostic test for major depressive disorder: a pilot and replication study. Mol Psychiatry. 2013;18:332–9.

Bjørnebekk A, Mathé AA, Brené S. The antidepressant effect of running is associated with increased hippocampal cell proliferation. Int J Neuropsychopharmacol. 2005;8:357–68.

Chapter 6

Koukopoulos A, Ghaemi SN. The primacy of mania: a reconsideration of mood disorders. Eur Psychiatry. 2009;24:125–34.

Phillips ML, Swartz HA. A critical appraisal of neuroimaging studies of bipolar disorder: toward a new conceptualization of underlying neural circuitry and a road map for future research. Am J Psychiatry. 2014;171:829–43.

Smith DJ, Griffiths E, Kelly M, Hood K, Craddock N, Simpson SA. Unrecognised bipolar disorder in primary care patients with depression. Br J Psychiatry. 2011;199:49–56.

Thompson JM, Gallagher P, Hughes JH, Watson S, Gray JM, Ferrier IN, et al. Neurocognitive impairment in euthymic patients with bipolar affective disorder. Br J Psychiatry. 2005;186:32–40.

Tordjman S, Drapier D, Bonnot O, Graignic R, Fortes S, Cohen D, et al. Animal models relevant to schizophrenia and autism: validity and limitations. Behav Genet. 2007;37:61–78.

Valet G. Cytomics as a new potential for drug discovery. Drug Discov Today. 2006;11:785–91.

Penninx BWJH, Beekman ATF, Smit JH, Zitman FG, Nolen WA, Spinhoven P, et al. The Netherlands Study of Depression and Anxiety (NESDA): rationale, objectives and methods. Int J Methods Psychiatr Res. 2008;17:121–40.

Bachmann RF, Schloesser RJ, Gould TD, Manji HK. Mood stabilizers target cellular plasticity and resilience cascades implications for the development of novel therapeutics. Mol Neurobiol. 2005;32(2):173–202.

Balanza-Martinez V, Fries GR, Colpo GD, Silveira PP, Portella AK, Tabarés-Seisdedos R, et al. Therapeutic use of omega-3 fatty acids in bipolar disorder. Expert Rev Neurother. 2011;11:1029–47.

Chiu C, Wang Z, Hunsberger J, Chuang D. Therapeutic potential of mood stabilizers lithium and valproic acid: beyond bipolar disorder. Pharmacol Rev. 2013;65:105–42.

Ebstein RP, Knafo A, Mankuta D, Chew SH, Lai PS. The contributions of oxytocin and vasopressin pathway genes to human behavior. Horm Behav. 2012;61:359–79.

Maes M, Goossens F, Scharpe S, Calabrese J, Desnyder R, Meltzer HY. Alterations in plasma prolylendopeptidase activity in depression, mania, and schizophrenia: effects of antidepressants, mood stabilizers, and antipsychotic drugs. Psychiatry Res. 1995;58:217–91.

Diazgranados N, Ibrahim L, Brutsche NE, Newberg A, Kronstein P, Khalife S, et al. A randomized add-on trial of an N-methyl-d-aspartate antagonist in treatment-resistant bipolar depression. Arch Gen Psychiatry. 2010;67:793–802.

Landreth G, Jiang Q, Mandrekar S, Heneka M. PPARgamma agonists as therapeutics for the treatment of Alzheimer's disease. Neurotherapeutics. 2008;5:481–9.

Haenisch F, Alsaif M, Guest PC, Rahmoune H, Yolken RH, Dickerson F, et al. Multiplex immunoassay analysis of plasma shows differences in biomarkers related to manic or mixed mood states in bipolar disorder patients. J Affect Disord. 2015;185:12–6.

Chapter 7

1. Barker DJ. Maternal Nutrition, Fetal Nutrition, and Disease in Later Life. Nutrition. 1997;13:807.

2. Hales CN, Barker DJ. Type 2 (non-insulin-dependent) diabetes mellitus: the thrifty phenotype hypothesis. Diabetologia. 1992;35:595–601.

3. Vaz RF, Mbajiorgu EF, Acuda SW. A preliminary study of stress levels among first year medical students at the University of Zimbabwe. Cent Afr J Med. 1998;44:214–9.
4. Charil A, Laplante DP, Vaillancourt C, King S. Prenatal stress and brain development. Brain research reviews. 2010;65:56–79.
5. Grizenko N, Shayan YR, Polotskaia A, Ter-Stepanian M, Joober R. Relation of maternal stress during pregnancy to symptom severity and response to treatment in children with ADHD. J Psychiatry Neurosci. 2008;33:10–6.
6. Susser E, Neugebauer R, Hoek HW, Brown AS, Lin S, Labovitz D, et al. Schizophrenia after prenatal famine. Further evidence. Arch Gen Psychiatry. 1996;53:25–31.
7. St Clair D, Xu M, Wang P, Yu Y, Fang Y, Zhang F, et al. Rates of adult schizophrenia following prenatal exposure to the Chinese famine of 1959-1961. JAMA. 2005;294:557–62.
8. Schmitz C, Rhodes ME, Bludau M, Kaplan S, Ong P, Ueffing I, et al. Depression: reduced number of granule cells in the hippocampus of female, but not male, rats due to prenatal restraint stress. Mol Psychiatry. 2009;7:810–3.
9. Ulupinar E, Yucel F. Prenatal stress reduces interneuronal connectivity in the rat cerebellar granular layer. Neurotoxicol Teratol. 2005;27:475–84.
10. Egaas B, Courchesne E, Saitoh O. Reduced size of corpus callosum in autism. Arch Neurol. 1995;52:794–801.
11. Seidman LJ, Valera EM, Makris N. Structural brain imaging of attention-deficit/hyperactivity disorder. Biol Psychiatry. 2005;57:1263–72.
12. King S, Laplante DP. The effects of prenatal maternal stress on children's cognitive development: Project Ice Storm. Stress. 2005;8:35–45.
13. King S, Laplante DP. Using natural disasters to study prenatal maternal stress in humans. Adv Neurobiol. 2015;10:285–313.
14. Heinrichs M, von Dawans B, Domes G. Oxytocin, vasopressin, and human social behavior. Front Neuroendocrinol. 2009;30:548–57.
15. Pariante CM, Vassilopoulou K, Velakoulis D, Phillips L, Soulsby B, Wood SJ, et al. Pituitary volume in psychosis. Br J Psychiatry. 2004;185:5–10.
16. Tannenbaum GS, Martin JB, Colle E. Ultradian growth hormone rhythm in the rat: effects of feeding, hyperglycemia, and insulin-induced hypoglycemia. Endocrinology. 1976;99:720–7.
17. Strous RD, Maayan R, Lapidus R, Stryjer R, Lustig M, Kotler M, et al. Dehydroepiandrosterone augmentation in the management of negative, depressive, and anxiety symptoms in schizophrenia. Arch Gen Psychiatry. 2003;60:133–41.
18. Gottschalk MG, Cooper JD, Chan MK, Bot M, Penninx BW, Bahn S. Discovery of serum biomarkers predicting development of a subsequent depressive episode in social anxiety disorder. Brain Behav Immun. 2015;48:123–31.

Chapter 8

Veenstra-Vanderweele J, Christian SL, Cook Jr EH. Autism as a paradigmatic complex genetic disorder. Annu Rev Genomics Hum Genet. 2004;5:379–405.
Bauman ML. Medical comorbidities in autism: challenges to diagnosis and treatment. Neurotherapeutics. 2010;7:320–7.
Baron-Cohen S. The extreme male brain theory of autism. Trends Cogn Sci. 2002;6:248–54.
Lord C, Rutter M, Le Couteur A. Autism diagnostic interview-revised: a revised version of a diagnostic interview for caregivers of individuals with possible pervasive developmental disorders. J Autism Dev Disord. 1994;24:659–85.

Sacco R, Curatolo P, Manzi B, Militerni R, Bravaccio C, Frolli A, et al. Principal pathogenetic components and biological endophenotypes in autism spectrum disorders. Autism Res. 2010;3:237–52.

Veenstra-VanderWeele J, Blakely RD. Networking in autism: leveraging genetic, biomarker and model system findings in the search for new treatments. Neuropsychopharmacology. 2012;37:196–212.

Lainhart JE, Piven J, Wzorek M, Landa R, Santangelo SL, Coon H, et al. Macrocephaly in children and adults with autism. J Am Acad Child Adolesc Psychiatry. 1997;36:282–90.

Kana RK, Libero LE, Hu CP, Deshpande HR, Colburn JS. Functional brain networks and white matter underlying theory-of-mind in Autism. Soc Cogn Affect Neurosci. 2014;9:98–105.

Miles JH. Autism spectrum disorders—a genetics review. Genet Med. 2011;13:278–94.

Voineagu I, Wang X, Johnston P, Lowe JK, Tian Y, Horvath S, et al. Transcriptomic analysis of autistic brain reveals convergent molecular pathology. Nature. 2011;474:380–4.

Corbett BA, Kantor AB, Schulman H, Walker WL, Lit L, Ashwood P, et al. A proteomic study of serum from children with autism showing differential expression of apolipoproteins and complement proteins. Mol Psychiatry. 2007;12:292–306.

Taurines R, Dudley E, Conner AC, Grassl J, Jans T, Guderian F, et al. Serum protein profiling and proteomics in autistic spectrum disorder using magnetic bead-assisted mass spectrometry. Eur Arch Psychiatry Clin Neurosci. 2010;260:249–55.

Schwarz E, Guest PC, Rahmoune H, Wang L, Levin Y, Ingudomnukul E, et al. Sex-specific serum biomarker patterns in adults with Asperger's syndrome. Mol Psychiatry. 2011;16:1213–20.

Broek JA, Guest PC, Rahmoune H, Bahn S. Proteomic analysis of post mortem brain tissue from autism patients: evidence for opposite changes in prefrontal cortex and cerebellum in synaptic connectivity-related proteins. Mol Autism. 2014;5:41.

Lake CR, Ziegler MG, Murphy DL. Increased norepinephrine levels and decreased dopamine-beta-hydroxylase activity in primary autism. Arch Gen Psychiatry. 1977;34:553–6.

Marshall CR, Noor A, Vincent JB, Lionel AC, Feuk L, Skaug J, et al. Structural variation of chromosomes in autism spectrum disorder. Am J Hum Genet. 2008;82:477–88.

Modahl C, Green L, Fein D, Morris M, Waterhouse L, Feinstein C, et al. Plasma oxytocin levels in autistic children. Biol Psychiatry. 1998;43:270–7.

Geier DA, Geier MR. A prospective assessment of androgen levels in patients with autistic spectrum disorders: biochemical underpinnings and suggested therapies. Neuro Endocrinol Lett. 2007;28:565–73.

Chapman E, Baron-Cohen S, Auyeung B, Knickmeyer R, Taylor K, Hackett G. Fetal testosterone and empathy: evidence from the empathy quotient (EQ) and the "reading the mind in the eyes" test. Soc Neurosci. 2006;1:135–48.

Singer HS, Morris CM, Williams PN, Yoon DY, Hong JJ, Zimmerman AW. Antibrain antibodies in children with autism and their unaffected siblings. J Neuroimmunol. 2006;178:149–55.

Tostes MHFS, Teixeira HC, Gattaz WF, Brandão MAF, Raposo NRB. Altered neurotrophin, neuropeptide, cytokines and nitric oxide levels in Autism. Pharmacopsychiatry. 2012;45:241–3.

Andersen IM, Kaczmarska J, McGrew SG, Malow BA. Melatonin for insomnia in children with autism spectrum disorders. J Child Neurol. 2008;23:482–5.

Bent S, Bertoglio K, Hendren RL. Omega-3 fatty acids for autistic spectrum disorder: a systemic review. J Autism Dev Disord. 2009;39:1145–54.

Oberman LM. mGluR antagonists and GABA agonists as novel pharmacological agents for the treatment of autism spectrum disorders. Expert Opin Investig Drugs. 2012;21:1819–25.

Miller G. Neuroscience. The promise and perils of oxytocin. Science. 2013;339:267–9.

Steeb H, Ramsey JM, Guest PC, Stocki P, Cooper JD, Rahmoune H, et al. Serum proteomic analysis identifies sex-specific differences in lipid metabolism and inflammation profiles in adults diagnosed with Asperger syndrome. Mol Autism. 2014;5:4.

Zheng Z, Li M, Lin Y, Ma Y. Effect of rosiglitazone on insulin resistance and hyperandrogenism in polycystic ovary syndrome. Zhonghua Fu Chan Ke Za Zhi. 2002;37:271–3.

Chapter 9

Czlonkowska A, Ciesielska A, Gromadzka G. Gender differences in neurological disease. Endocrine. 2006;29:243–56.

Tingen CM, Kim AM, Wu P-H, Woodruff TK. Sex and sensitivity: the continued need for sex-based biomedical research and implementation. Women's Health. 2010;6:511–6.

Eskes T, Haanen C. Why do women live longer than men? Eur J Obst Gyn Reprod Biol. 2007;133:126–33.

Lee E, Beaver KM, Wright J. Handbook of crime correlates. Academic Press, 2009. ISBN: 0123736129.

Ramsey JM, Schwarz E, Guest PC, van Beveren NJ, Leweke FM, Rothermundt M, et al. Molecular sex differences in human serum. PLoS One. 2012;7, e51504.

Ingalhalikar M, Smith A, Parker D, Satterthwaite TD, Elliott MA, Ruparel K, et al. Sex differences in the structural connectome of the human brain. Proc Natl Acad Sci U S A. 2013;111:823–8.

Speck O, Ernst T, Braun J, Koch C, Miller E, Chang L. Gender differences in the functional organization of the brain for working memory. Neuroreport. 2000;11:2581–5.

Schmitt DP, Realo A, Voracek M, Allik J. Why can't a man be more like a woman? Sex differences in Big Five personality traits across 55 cultures. J Pers SocPsychol. 2008;94:168–82.

Ochoa S, Usall J, Cobo J, Labad X, Kulkarni J. Gender differences in schizophrenia and first-episode psychosis: a comprehensive literature review. Schizophr Res Treatment. 2012;2012:916198.

Goldstein JM, Santangelo SL, Simpson JC, Tsuang MT. The role of gender in identifying subtypes of schizophrenia: a latent class analytic approach. Schizophrenia Bull. 1990;16:263–75.

Ramsey JM, Schwarz E, Guest PC, van Beveren NJ, Leweke FM, Rothermundt M, et al. Distinct molecular phenotypes in male and female schizophrenia patients. PLoS One. 2013;8, e78729.

Schuch JJ, Roest AM, Nolen WA, Penninx BW, de Jonge P. Gender differences in major depressive disorder: results from the Netherlands study of depression and anxiety. J Affect Disord. 2014;156:156–63.

Soares CN, Almeida OP, Joffe H, Cohen LS. Efficacy of estradiol for the treatment of depressive disorders in perimenopausal women: a double-blind, randomized, placebo-controlled trial. Arch Gen Psychiatry. 2001;58:529–34.

Khan A, Brodhead AE, Schwartz KA, Kolts RL, Brown WA. Sex differences in antidepressant response in recent antidepressant clinical trials. J Clin Psychopharmacol. 2005;25:318–24.

Angold A, Worthman CS. Puberty onset of gender differences in rates of depression: a developmental, epidemiologic and neuroendocrine perspective. J Affect Disord. 1993;29:145–58.

Arnold AP, Breedlove SM. Organizational and activational effects of sex steroids on brain and behavior: a reanalysis. Horm Behav. 1985;19:469–98.

Barontini M, Garcia-Rudaz MC, Veldhuis JD. Mechanisms of hypothalamic-pituitary-gonadal disruption in polycystic ovarian syndrome. Arch Med Res. 2001;32:544–52.

Zheng Z, Li M, Lin Y, Ma Y. Effect of rosiglitazone on insulin resistance and hyperandrogenism in polycystic ovary syndrome. Zhonghua Fu Chan Ke Za Zhi. 2002;37:271–3.

Chapter 10

Kalaria RN, Maestre GE, Arizaga R, Friedland RP, Galasko D, Hall K, et al. Alzheimer's disease and vascular dementia in developing countries: prevalence, management, and risk factors. Lancet Neurol. 2008;7:812–26.

http://www.who.int/mediacentre/factsheets/fs362/en/

Rentz DM, Parra Rodriguez MA, Amariglio R, Stern Y, Sperling R, Ferris S. Promising developments in neuropsychological approaches for the detection of preclinical Alzheimer's disease: a selective review. Alzheimers Res Ther. 2013;5:58.

Ono M, Saji H. Molecular approaches to the treatment, prophylaxis, and diagnosis of Alzheimer's disease: novel PET/SPECT imaging probes for diagnosis of Alzheimer's disease. J Pharmacol Sci. 2012;118:338–44.

Zafari S, Backes C, Meese E, Keller A. Circulating biomarker panels in Alzheimer's disease. Gerontology. 2015;61:497–503.

Trushina E, Mielke MM. Recent advances in the application of metabolomics to Alzheimer's Disease. Biochim Biophys Acta. 1842;2014:1232–9.

Barron AM, Pike CJ. Sex hormones, aging, and Alzheimer's disease. Front Biosci (Elite Ed). 2012;4:976–97.

Venigalla M, Sonego S, Gyengesi E, Sharman MJ, Münch G. Novel promising therapeutics against chronic neuroinflammation and neurodegeneration in Alzheimer's disease. Neurochem Int. 2016;95:63–74.

Correia SC, Santos RX, Perry G, Zhu X, Moreira PI, Smith MA. Insulin-resistant brain state: the culprit in sporadic Alzheimer's disease? Ageing Res Rev. 2011;10:264–73.

Jia Q, Deng Y, Qing H. Potential therapeutic strategies for Alzheimer's disease targeting or beyond β-amyloid: insights from clinical trials. Biomed Res Int. 2014;2014:837157.

Sabharwal P, Wisniewski T. Novel immunological approaches for the treatment of Alzheimer's disease. Zhongguo Xian Dai Shen Jing Ji Bing Za Zhi. 2014;14:139–51.

Liu QP, Wu YF, Cheng HY, Xia T, Ding H, Wang H, et al. Habitual coffee consumption and risk of cognitive decline/dementia: a systematic review and meta-analysis of prospective cohort studies. Nutrition. 2016;32:628–36.

Thaipisuttikul P, Galvin JE. Use of medical foods and nutritional approaches in the treatment of Alzheimer's disease. Clin Pract (Lond). 2012;9:199–209.

Farina N, Rusted J, Tabet N. The effect of exercise interventions on cognitive outcome in Alzheimer's disease: a systematic review. Int Psychogeriatr. 2014;26:9–18.

Bertram S, Brixius K, Brinkmann C. Exercise for the diabetic brain: how physical training may help prevent dementia and Alzheimer's disease in T2DM patients. Endocrine. 2016 May;9 [Epub ahead of print].

Chapter 11

LeWitt PA. Levodopa for the treatment of Parkinson's disease. N Engl J Med. 2008;359:2468–76.

Dibble LE, Cavanaugh JT, Earhart GM, Ellis TD, Ford MP, Foreman KB. Charting the progression of disability in Parkinson disease: study protocol for a prospective longitudinal cohort study. BMC Neurol. 2010;10:110.

Poewe W, Mahlknecht P, Jankovic J. Emerging therapies for Parkinson's disease. Curr Opin Neurol. 2012;25:448–59.

Trenkwalder C, Chaudhuri KR, García Ruiz PJ, LeWitt P, Katzenschlager R, Sixel-Döring F, et al. Expert Consensus Group report on the use of apomorphine in the treatment of Parkinson's disease–Clinical practice recommendations. Parkinsonism Relat Disord. 2015;21:1023–30.

Kalia LV, Kalia SK, Lang AE. Disease-modifying strategies for Parkinson's disease. Mov Disord. 2015;30:1442–50.

Chahine LM, Stern MB. Diagnostic markers for Parkinson's disease. Curr Opin Neurol. 2011;24:309–17.

Miller DB, O'Callaghan JP. Biomarkers of Parkinson's disease: present and future. Metabolism. 2015;64(3 Suppl 1):S40–6.

Lotia M, Jankovic J. New and emerging medical therapies in Parkinson's disease. Expert Opin Pharmacother. 2016;17:895–909.

Foulds PG, Diggle P, Mitchell JD, Parker A, Hasegawa M, Masuda-Suzukake M, et al. A longitudinal study on α-synuclein in blood plasma as a biomarker for Parkinson's disease. Sci Rep. 2013;3:2540.

https://www.barrowneuro.org/get-to-know-barrow/centers-programs/muhammad-ali-parkinson-center/

Chapter 12

Mitchell AJ, Vaze A, Rao S. Clinical diagnosis of depression in primary care: a meta-analysis. Lancet. 2009;374:609–19.

Bet PM, Hugtenburg JG, Penninx BW, Av B, Nolen WA, Hoogendijk WJ. Treatment inadequacy in primary and specialized care patients with depressive and/or anxiety disorders. Psychiatry Res. 2013;210:594–600.

Anderson NL, Anderson NG. The human plasma proteome: history, character, and diagnostic prospects. Mol Cell Proteomics. 2002;1:845–67.

Schumacher S, Nestler J, Otto T, Wegener M, Ehrentreich-Förster E, Michel D, et al. Highly-integrated lab-on-chip system for point-of-care multiparameter analysis. Lab Chip. 2012;12:464–73.

Mental well-being: For a smart, inclusive and sustainable Europe. EC Report, 2011.ec.europa.eu/health/mental_health/docs/outcomes_pact_en.pdf

Fu Q, Jie Zhu J, Van Eyk JE. Comparison of multiplex immunoassay platforms. Clin Chem. 2010;56:314–8.

Lee H, Xu L, Koh D, Nyayapathi N, Oh KW. Various on-chip sensors with microfluidics for biological applications. Sensors (Basel). 2014;14:17008–36.

Cosnier S. Immobilization of biomolecules by electropolymerised films. In: Marks RS, Cullen DC, Karube I, Lowe CR, Weetall H, editors. Handbook of biosensors and biochips. Sussex: Wiley Interscience; 2007. p. 237–49.

Holzinger M, Le Goff A, Cosnier S. Nanomaterials for biosensing applications: a review. Front Chem. 2014;2:63.

Li PCH. Fundamentals of microfluidics and lab on a chip for biological analysis and discovery. Boca Raton, FL: CRC Press/Taylor & Francis Group; 2010.

Klasnja P, Pratt W. Healthcare in the pocket: mapping the space of mobile-phone health interventions. J Biomed Inform. 2012;45:184–98.

Berg B, Cortazar B, Tseng D, Ozkan H, Feng S, Wei Q, et al. Cellphone-based hand-held microplate reader for point-of-care testing of enzyme-linked immunosorbent assays. ACS Nano. 2015.

Index

© Springer International Publishing AG 2017
P.C. Guest, *Biomarkers and Mental Illness*,
DOI 10.1007/978-3-319-46088-8

Printed in the United States
By Bookmasters